Pitman Research Notes in Mathematics Series

Submission of proposals for consideration

Suggestions for publication, in the form of outlines and representative samples, are invited by the Editorial Board for assessment. Intending authors should approach one of the main editors or another member of the Editorial Board, citing the relevant AMS subject classifications. Alternatively, outlines may be sent directly to the publisher's offices. Refereeing is by members of the board and other mathematical authorities in the topic concerned, throughout the world.

Preparation of accepted manuscripts

On acceptance of a proposal, the publisher will supply full instructions for the preparation of manuscripts in a form suitable for direct photo-lithographic reproduction. Specially printed grid sheets can be provided and a contribution is offered by the publisher towards the cost of typing. Word processor output, subject to the publisher's approval, is also acceptable.

Illustrations should be prepared by the authors, ready for direct reproduction without further improvement. The use of hand-drawn symbols should be avoided wherever possible, in order to maintain maximum clarity of the text.

The publisher will be pleased to give any guidance necessary during the preparation of a typescript, and will be happy to answer any queries.

Important note

In order to avoid later retyping, intending authors are strongly urged not to begin final preparation of a typescript before receiving the publisher's guidelines. In this way it is hoped to preserve the uniform appearance of the series.

Addison Wesley Longman Ltd
Edinburgh Gate
Harlow, Essex, CM20 2JE
UK
(Telephone (0) 1279 623623)

Titles in this series. A full list is available from the publisher on request.

251 Stability of stochastic differential equations with respect to semimartingales
X Mao
252 Fixed point theory and applications
J Baillon and M Théra
253 Nonlinear hyperbolic equations and field theory
M K V Murthy and S Spagnolo
254 Ordinary and partial differential equations. Volume III
B D Sleeman and R J Jarvis
255 Harmonic maps into homogeneous spaces
M Black
256 Boundary value and initial value problems in complex analysis: studies in complex analysis and its applications to PDEs 1
R Kühnau and W Tutschke
257 Geometric function theory and applications of complex analysis in mechanics: studies in complex analysis and its applications to PDEs 2
R Kühnau and W Tutschke
258 The development of statistics: recent contributions from China
X R Chen, K T Fang and C C Yang
259 Multiplication of distributions and applications to partial differential equations
M Oberguggenberger
260 Numerical analysis 1991
D F Griffiths and G A Watson
261 Schur's algorithm and several applications
M Bakonyi and T Constantinescu
262 Partial differential equations with complex analysis
H Begehr and A Jeffrey
263 Partial differential equations with real analysis
H Begehr and A Jeffrey
264 Solvability and bifurcations of nonlinear equations
P Drábek
265 Orientational averaging in mechanics of solids
A Lagzdins, V Tamuzs, G Teters and A Kregers
266 Progress in partial differential equations: elliptic and parabolic problems
C Bandle, J Bemelmans, M Chipot, M Grüter and J Saint Jean Paulin
267 Progress in partial differential equations: calculus of variations, applications
C Bandle, J Bemelmans, M Chipot, M Grüter and J Saint Jean Paulin
268 Stochastic partial differential equations and applications
G Da Prato and L Tubaro
269 Partial differential equations and related subjects
M Miranda
270 Operator algebras and topology
W B Arveson, A S Mishchenko, M Putinar, M A Rieffel and S Stratila
271 Operator algebras and operator theory
W B Arveson, A S Mishchenko, M Putinar, M A Rieffel and S Stratila
272 Ordinary and delay differential equations
J Wiener and J K Hale
273 Partial differential equations
J Wiener and J K Hale
274 Mathematical topics in fluid mechanics
J F Rodrigues and A Sequeira
275 Green functions for second order parabolic integro-differential problems
M G Garroni and J F Menaldi
276 Riemann waves and their applications
M W Kalinowski
277 Banach C(K)-modules and operators preserving disjointness
Y A Abramovich, E L Arenson and A K Kitover
278 Limit algebras: an introduction to subalgebras of C*-algebras
S C Power
279 Abstract evolution equations, periodic problems and applications
D Daners and P Koch Medina
280 Emerging applications in free boundary problems
J Chadam and H Rasmussen
281 Free boundary problems involving solids
J Chadam and H Rasmussen
282 Free boundary problems in fluid flow with applications
J Chadam and H Rasmussen
283 Asymptotic problems in probability theory: stochastic models and diffusions on fractals
K D Elworthy and N Ikeda
284 Asymptotic problems in probability theory: Wiener functionals and asymptotics
K D Elworthy and N Ikeda
285 Dynamical systems
R Bamon, R Labarca, J Lewowicz and J Palis
286 Models of hysteresis
A Visintin
287 Moments in probability and approximation theory
G A Anastassiou
288 Mathematical aspects of penetrative convection
B Straughan
289 Ordinary and partial differential equations. Volume IV
B D Sleeman and R J Jarvis
290 *K*-theory for real *C**-algebras
H Schröder
291 Recent developments in theoretical fluid mechanics
G P Galdi and J Necas
292 Propagation of a curved shock and nonlinear ray theory
P Prasad
293 Non-classical elastic solids
M Ciarletta and D Ieşan
294 Multigrid methods
J Bramble
295 Entropy and partial differential equations
W A Day
296 Progress in partial differential equations: the Me surveys 2
M Chipot
297 Nonstandard methods in the calculus of variatio
C Tuckey
298 Barrelledness, Baire-like- and (LF)-spaces
M Kunzinger
299 Nonlinear partial differential equations and thei applications. Collège de France Seminar. Volume XI
H Brezis and J L Lions

300 Introduction to operator theory
T Yoshino
301 Generalized fractional calculus and applications
V Kiryakova
302 Nonlinear partial differential equations and their applications. Collège de France Seminar Volume XII
H Brezis and J L Lions
303 Numerical analysis 1993
D F Griffiths and G A Watson
304 Topics in abstract differential equations
S Zaidman
305 Complex analysis and its applications
C C Yang, G C Wen, K Y Li and Y M Chiang
306 Computational methods for fluid-structure interaction
J M Crolet and R Ohayon
307 Random geometrically graph directed self-similar multifractals
L Olsen
308 Progress in theoretical and computational fluid mechanics
G P Galdi, J Málek and J Necas
309 Variational methods in Lorentzian geometry
A Masiello
310 Stochastic analysis on infinite dimensional spaces
H Kunita and H-H Kuo
311 Representations of Lie groups and quantum groups
V Baldoni and M Picardello
312 Common zeros of polynomials in several variables and higher dimensional quadrature
Y Xu
313 Extending modules
N V Dung, D van Huynh, P F Smith and R Wisbauer
314 Progress in partial differential equations: the Metz surveys 3
M Chipot, J Saint Jean Paulin and I Shafrir
315 Refined large deviation limit theorems
V Vinogradov
316 Topological vector spaces, algebras and related areas
A Lau and I Tweddle
317 Integral methods in science and engineering
C Constanda
318 A method for computing unsteady flows in porous media
R Raghavan and E Ozkan
319 Asymptotic theories for plates and shells
R P Gilbert and K Hackl
320 Nonlinear variational problems and partial differential equations
A Marino and M K V Murthy
321 Topics in abstract differential equations II
S Zaidman
322 Diffraction by wedges
B Budaev
323 Free boundary problems: theory and applications
J I Diaz, M A Herrero, A Liñan and J L Vazquez
324 Recent developments in evolution equations
A C McBride and G F Roach
325 Elliptic and parabolic problems: Pont-à-Mousson 1994
C Bandle, J Bemelmans, M Chipot, J Saint Jean Paulin and I Shafrir
326 Calculus of variations, applications and computations: Pont-à-Mousson 1994
C Bandle, J Bemelmans, M Chipot, J Saint Jean Paulin and I Shafrir
327 Conjugate gradient type methods for ill-posed problems
M Hanke
328 A survey of preconditioned iterative methods
A M Bruaset
329 A generalized Taylor's formula for functions of several variables and certain of its applications
J-A Riestra
330 Semigroups of operators and spectral theory
S Kantorovitz
331 Boundary-field equation methods for a class of nonlinear problems
G N Gatica and G C Hsiao
332 Metrizable barrelled spaces
J C Ferrando, M López Pellicer and L M Sánchez Ruiz
333 Real and complex singularities
W L Marar
334 Hyperbolic sets, shadowing and persistence for noninvertible mappings in Banach spaces
B Lani-Wayda
335 Nonlinear dynamics and pattern formation in the natural environment
A Doelman and A van Harten
336 Developments in nonstandard mathematics
N J Cutland, V Neves, F Oliveira and J Sousa-Pinto
337 Topological circle planes and topological quadrangles
A E Schroth
338 Graph dynamics
E Prisner
339 Localization and sheaves: a relative point of view
P Jara, A Verschoren and C Vidal
340 Mathematical problems in semiconductor physics
P Marcati, P A Markowich and R Natalini
341 Surveying a dynamical system: a study of the Gray–Scott reaction in a two-phase reactor
K Alhumaizi and R Aris
342 Solution sets of differential equations in abstract spaces
R Dragoni, J W Macki, P Nistri and P Zecca
343 Nonlinear partial differential equations
A Benkirane and J-P Gossez
344 Numerical analysis 1995
D F Griffiths and G A Watson
345 Progress in partial differential equations: the Metz surveys 4
M Chipot and I Shafrir
346 Rings and radicals
B J Gardner, Liu Shaoxue and R Wiegandt
347 Complex analysis, harmonic analysis and applications
R Deville, J Esterle, V Petkov, A Sebbar and A Yger
348 The theory of quantaloids
K I Rosenthal
349 General theory of partial differential equations and microlocal analysis
Qi Min-you and L Rodino

350 Progress in elliptic and parabolic partial differential equations
A Alvino, P Buonocore, V Ferone, E Giarrusso, S Matarasso, R Toscano and G Trombetti
351 Integral representations for spatial models of mathematical physics
V V Kravchenko and M V Shapiro
352 Dynamics of nonlinear waves in dissipative systems: reduction, bifurcation and stability
G Dangelmayr, B Fiedler, K Kirchgässner and A Mielke
353 Singularities of solutions of second order quasilinear equations
L Véron
354 Mathematical theory in fluid mechanics
G P Galdi, J Málek and J Necas
355 Eigenfunction expansions, operator algebras and symmetric spaces
R M Kauffman
356 Lectures on bifurcations, dynamics and symmetry
M Field
357 Noncoercive variational problems and related results
D Goeleven
358 Generalised optimal stopping problems and financial markets
D Wong
359 Topics in pseudo-differential operators
S Zaidman

S Zaidman

University of Montreal, Canada

Topics in pseudo-differential operators

LONGMAN

Addison Wesley Longman Limited
Edinburgh Gate, Harlow
Essex CM20 2JE, England
and Associated Companies throughout the world.

*Published in the United States of America
by Addison Wesley Longman Inc.*

First published 1996

AMS Subject Classifications: (Main) 35, 46, 47
(Subsidiary) 35S05, 46F12, 47G30

ISSN 0269-3674

ISBN 0 582 27782 5

British Library Cataloguing in Publication Data

A catalogue record for this book is
available from the British Library

Library of Congress Cataloging-in-Publication data

Zaidman, Samuel, 1933-
Topics in pseudo-differential operators / S. Zaidman.
p. cm. -- (Pitman research notes in mathematics series, ISSN 0269-3674 ; ??)
ISBN 0-582-27782-5 (alk. paper)
1. Pseudodifferential operators. I. Title. II. Series.
QA329.7.Z35 1996
515'.7242--dc20 96-31181
CIP

Printed and bound in Great Britain
by Biddles Ltd, Guildford and King's Lynn

Contents

Introduction

I	Order and true order of linear operators in some vector spaces	1
II	Asymptotic expansions of linear operators in some vector spaces	17
III	Pseudo-differential operators in the spaces $\mathcal{B}_{1,s}(\mathbb{R}^n)$	27
IV	Pseudo-differential operators in $\mathcal{F}^{-1}(L^p)$ and in $H^s(\mathbb{R}^n)$ spaces	40
V	Alternative representation formulas for operators $G(x,D)$ and $\mathcal{G}(x,D)$	46
VI	Kohn-Nirenberg homogeneous and C^∞-symbols and their associated operators	50
VII	Compactness of the operator $A(x,D) - \mathcal{A}(x,D)$ in the space $\mathcal{F}^{-1}(L^1(\mathbb{R}^n))(A(x,D) \equiv \mathcal{A}(x,D)$-modulo the compact operators)	66
VIII	Gohberg's Lemma and applications	79
	References	110
	Index of symbols	112
	Subject index	116

Introduction

The present monograph is, essentially, a continuation of our book "Distributions and Pseudo-differential Operators", [9].

Again, as in [9], we present definitions and results pertaining to pseudo-differential operators as they were initially considered by Kohn and Nirenberg [5]; extensions and variants of some parts of [5] appear in the following pages.

We made usage of the important monograph by Friedrichs [4] and the lecture notes by Andreotti–Spagnolo [1].

Part of the results below appeared – more or less in the same form – in some articles of myself ([10], [11], [12], [13], as listed in the page on "References"). They were also presented in Seminars at the Banaras Hindu University, I-III-1992. We hope that the unified presentation in this volume will help the understanding of some nice aspects of Kohn-Nirenberg's theory of pseudo-differential operators.

Chapter I
Order and true order of linear operators in some vector spaces

Introduction

The concepts of order and true order of linear operators in the space $\mathcal{S}(\mathbb{R}^n)$ of C^∞ – rapidly decreasing functions were introduced by Kohn-Nirenberg in [5]; discussions and generalizations appear in [1]–[2].

In the present Chapter we explain – following our article [13] – a somewhat different concept of order and true order which can be applied to linear operators acting on linear subspaces of the intersection of a scale of Banach spaces; we establish some simple properties and give a few "concrete" examples.

1. First definitions and examples

We consider a family of Banach spaces E^s, over the same (real or complex) field K, depending on the real parameter s, $-\infty < s < +\infty$. We assume the following:

i) E^s is a vector subspace of $E^{s'}$ for $s \geq s'$

$$\tag{1.1}$$

ii) The inequalities

$$\| u \|_{E^{s'}} \leq \| u \|_{E^s} \ \forall u \in E^s, \text{ if } s' \leq s \tag{1.2}$$

hold true.

A family $\{E^s\}$ as above will be called a *scale* of Banach spaces.

A trivial example : take a single Banach space E, and put then $E^s = E$ for all $s \in \mathbb{R}$ (the real line).

Some classical (nontrivial) examples: (a) The $H^s(\mathbb{R}^n)$ spaces, $(s \in \mathbb{R})$, consisting of temperate distributions T such that, if $\hat{T}$ is the Fourier transform, then $(1+|\xi|^2)^{s/2}\hat{T}(\xi) \in L^2(\mathbb{R}^n)$. Then the family $\{H^s(\mathbb{R}^n)\}_{s\in\mathbb{R}}$ form a scale of (Hilbert) spaces over $\mathbb{C}$ (the complex field) with the $\| \ \|_s$ given by relation

$$\| T \|_{H^s} = \left(\int_{\mathbb{R}^n} (1+|\xi|^2)^s |\hat{T}(\xi)|^2 d\xi \right)^{1/2} \tag{1.3}$$

(see our monograph [9] – Ch. 6).

(b) The $\mathcal{B}_{1,s}(\mathbb{R}^n)$ spaces – consisting of those temperate distributions T such that $(1+|\xi|^2)^{s/2}\hat{T}(\xi) \in L^1(\mathbb{R}^n)$, (see [9] – Ch. 12). The family $\{\mathcal{B}_{1,s}(\mathbb{R}^n)\}_{s\in\mathbb{R}}$ is a scale of (Banach) spaces over $\mathbb{C}$, where the norm in $\mathcal{B}_{1,s}$ is given by the formula

$$\| T \|_{\mathcal{B}_{1,s}} = \int_{\mathbb{R}^n} (1+|\xi|^2)^{s/2} |\hat{T}(\xi)| d\xi \tag{1.4}$$

Next, given a general scale of Banach spaces $\{E^s\}_{s\in\mathbb{R}}$ we define a space E^∞ as the (set theoretic) intersection of all of E^s. Thus

$$E^\infty = \bigcap_{s\in\mathbb{R}} E^s = \bigcap_{s\geq 0} E^s = \bigcap_{k\in\mathbb{N}} E^k \tag{1.5}$$

The equalities in (1.5) are a direct consequence of (1.1) (for instance, let $x \in E^k \forall k \in \mathbb{N}$; then $x \in E^s$ for $0 \leq s \leq k$, and this $\forall k \in \mathbb{N}$).

We readily see that E^∞ is (also) a vector space over the field K. Our next step:

we consider an arbitrary vector subspace of E^∞, denoted here with V

(in above example (a), E^∞ is the (well-known) space $H^\infty(\mathbb{R}^n)$ and we can take $V = \mathcal{S}(\mathbb{R}^n)$ (the Schwartz space of $C^\infty(\mathbb{R}^n)$ – rapidly decreasing functions); in example (b) we have $E^\infty = \mathcal{B}_{1,\infty}(\mathbb{R}^n) = \bigcap_{s\in\mathbb{R}} \mathcal{B}_{1,s}(\mathbb{R}^n)$, and again we note that $\mathcal{S}(\mathbb{R}^n)$ is a vector subspace of $\mathcal{B}_{1,\infty}(\mathbb{R}^n)$).

We now consider the vector space $\mathcal{L}in(V)$; its elements are all the linear operators, $V \to V$; they form an algebra over K.

Extending a definition from Kohn-Nirenberg [5] (they take the scale $\{H^s(\mathbb{R}^n)\}$ and $V = \mathcal{S}(\mathbb{R}^n)$) we say that

an element L of $\mathcal{L}in(V)$ has order r or is of order r when for every real number s there is a constant $C_{s,r} \geq 0$, such that

$$\| Lu \|_{E^s} \leq C_{s,r} \| u \|_{E^{s+r}}, \ \forall u \in V \tag{1.6}$$

holds true.

Remark It is clear that if the operator L has order r it also has order r' for any $r' \in \mathbb{R}, r' > r$.

We also say that

The infimum (G.L.B.) of all orders r of the operator L is (by definition), the true order of $L : t.o(L) = \inf\{r \in \mathbb{R}, L \text{ is of order } r\}$.

It is also clear that if operator L has order r and the operator M has order s, then the product operators $L \circ M$ or $M \circ L$ have order $r + s$.

The set $\{r \in \mathbb{R}, L$ is of order $r\}$ will be called order set of L. Thus, for any $L \in \mathcal{L}in(V)$, where V is a linear subspace of E^{∞}, the order set of L, denoted $\mathcal{O}(L)$, is the subset of real numbers defined by $\mathcal{O}(L) = \{r \in \mathbb{R}$, such that, $\forall s \in \mathbb{R}, \exists C_{s,r} > 0$, with property that

$$\| Lu \|_{E^s} \leq C_{s,r} \| u \|_{E^{s+r}}, \ \forall u \in V\} \tag{1.7}$$

Accordingly, the mapping $L \to \mathcal{O}(L)$ is a map from $\mathcal{L}in(V)$ in $\mathcal{P}(\mathbb{R}) = 2^{\mathbb{R}}$ (the power set of $\mathbb{R}$), and if $r_1 \in \mathcal{O}(L), r_2 > r_1$, then $r_2 \in \mathcal{O}(L)$. Again, we see that

$$t.o(L) = \inf \mathcal{O}(L) \tag{1.8}$$

Let us note the following

Proposition 1.1 *Let be $L \in \mathcal{L}in(V)$ and $r \in \mathcal{O}(L)$. Take a fixed $s \in \mathbb{R}$ and assume that V is dense in E^{s+r} (in $(s+r)$-norm). Then, the operator L can be extended to a linear continuous operator, $E^{s+r} \to E^s$, whose operator norm is $\leq C_{s,r}$.*

Proof The operator $L, V \to V$, in view of (1.7) is Lipschitz continuous: $\| Lu - Lv \|_{E^s} \leq C_{s,r} \| u - v \|_{E^{s+r}}, \ \forall u, v \in V$. Therefore, if $(u_n)_1^{\infty}$ is a Cauchy sequence of elements of V in E^{s+r}-norm, then $(Lu_n)_1^{\infty}$ is also a Cauchy sequence in the E^s-norm. This permits, in the usual well-known way, to extend L from V to the whole of E^{s+r}: if $\nu \in E^{s+r}$ and (u_n) is a sequence in V such that $\| u_n - \nu \|_{E^{s+r}} \to 0$, then we define $\tilde{L}\nu = \lim_{n\to\infty} Lu_n$, and $\tilde{L}\nu \in E^s$. Moreover, we get the estimate

$$\| \tilde{L}\nu \|_{E^s} = \lim_{n\to\infty} \| Lu_n \|_{E^s} \leq C_{s,r} \lim_{n\to\infty} \| u_n \|_{E^{s+r}} = C_{s,r} \| \nu \|_{E^{s+r}} \tag{1.9}$$

which says that $\tilde{L} \in \mathcal{L}(E^{s+r}, E^s)$ and $\| \tilde{L} \| \leq C_{s,r}$.

We shall now discuss a few "concrete" examples:

a) Consider a bounded measurable function $\varphi(\xi), \mathbb{R}^n \to \mathbb{C}$ and then take $u \in H^{\infty}(\mathbb{R}^n)$; thus $(1 + |\xi|^2)^{s/2}\hat{u}(\xi) \in L^2(\mathbb{R}^n), \ \forall s \in \mathbb{R}$. It follows that $(1 + |\xi|^2)^{s/2}\varphi(\xi)\hat{u}(\xi)$ also belongs to $L^2(\mathbb{R}^n)$ $\forall s \in \mathbb{R}$. Then we see that $\varphi(\xi)\hat{u}(\xi) \in L^1_{loc}(\mathbb{R}^n) \cap L^2(\mathbb{R}^n)$. Accordingly, the inverse Fourier transform (in $\mathcal{S}'(\mathbb{R}^n)$ sense):

$\mathcal{F}^{-1}[\varphi(\xi)\hat{u}(\xi)]$ will exist and will belong to $H^s(\mathbb{R}^n)$ for all $s \in \mathbb{R}$. Define accordingly an operator $\varphi(D)$ on H^∞ via the formula

$$\varphi(D)u = \mathcal{F}^{-1}[\varphi(\xi)(\mathcal{F}u(\xi)], \ \forall u \in H^\infty \tag{1.10}$$

or also

$$\varphi(D) = \mathcal{F}^{-1}\mathcal{M}_{\varphi(\cdot)}\mathcal{F} \tag{1.11}$$

where $\mathcal{M}_{\varphi(\cdot)}$ is the multiplication operator by $\varphi(\cdot)$.

Therefore, we have : $\varphi(D)$ maps H^∞ into itself; furthermore :

$$\| \varphi(D)u \|_{H^s} = \left(\int_{\mathbb{R}^n} (1+|\xi|^2)^s |\varphi(\xi)|^2 |\hat{u}(\xi)|^2 d\xi \right)^{1/2} \leq \underset{\mathbb{R}^n}{ess.sup} |\varphi(\xi)|. \| u \|_{H^s} \tag{1.12}$$

It follows that the order set $\mathcal{O}(\varphi(D))$ contains the interval $[0, +\infty)$.

In a slightly more general situation, let us take a measurable function $\varphi(\xi), \mathbb{R}^n \to \mathbb{C}$, which satisfies an estimate

$$|\varphi(\xi)| \leq C(1+|\xi|^2)^\sigma, \ \forall \xi \in \mathbb{R}^n \text{ (for some real number } \sigma) \tag{1.13}$$

It follows that, again, the operator $\varphi(D) = \mathcal{F}^{-1}\mathcal{M}_{\varphi(\cdot)}\mathcal{F}$ is well defined as a linear operator, $H^\infty \to H^\infty$ and estimates

$$\| \varphi(D)u \|_{H^s} \leq C \| u \|_{H^{s+2\sigma}}, \ \forall s \in \mathbb{R}, \ \forall u \in H^\infty \tag{1.14}$$

hold true. Accordingly, $2\sigma \in \mathcal{O}(\varphi(D))$ and $[2\sigma, +\infty) \subset \mathcal{O}(\varphi(D))$.

An important special case is obtained with a continuous complex-valued function on $\mathbb{R}^n, \varphi(\xi)$, which vanishes outside some compact subset of $\mathbb{R}^n$. It is then obvious that a sequence of estimates

$$|\varphi(\xi)| \leq C_p(1+|\xi|^2)^{-p}, \ \forall p \in \mathbb{N}, \ \forall \xi \in \mathbb{R}^n \tag{1.15}$$

hold true. Accordingly, the corresponding operator $H^\infty \to H^\infty$ has order $-2p \ \forall p \in \mathbb{N}$. The order set $\mathcal{O}(\varphi(D))$ is now the whole real line and $t.o(\varphi(D)) = -\infty$.

(b) Next we refer to the monograph [9]. Consider "symbols" $p(x, \xi) \in S^r$ and the (pseudo-differential) operator associated to them (page 94) (therefore

$p(x,\xi), \mathbb{R}^n \times \mathbb{R}^n \to \mathbb{C}$ belongs to $C^\infty(\mathbb{R}^n \times \mathbb{R}^n)$ and $\forall \ell \in \underline{\mathbb{N}}, \alpha \in \underline{\mathbb{N}}^n, \beta \in \underline{\mathbb{N}}^n, \exists C_{\alpha,\beta,\ell} > 0$ such that the estimates

$$(1+|x|^2)^\ell |\partial_\xi^\alpha \partial_x^\beta p(x,\xi)| \leq C_{\alpha,\beta,\ell}\,(1+|\xi|)^{r-|\alpha|},\ \forall (x,\xi) \in \mathbb{R}^n \times \mathbb{R}^n \qquad (1.16)$$

are satisfied (with a fixed real number r); the operator $\mathcal{P}(x,D)$ acting on $\mathcal{S}(\mathbb{R}^n)$ is defined by the formula

$$(\mathcal{P}(x,D)u)(x) = (2\pi)^{-n/2} \int_{\mathbb{R}^n} e^{i<x,\xi>} p(x,\xi)\hat{u}(\xi)d\xi,\ \forall u \in \mathcal{S}(\mathbb{R}^n) \qquad (1.17)$$

and, as seen in [9], it maps (linearly), $\mathcal{S}(\mathbb{R}^n)$ into $C^\infty(\mathbb{R}^n)$.

We shall see that, actually, $\mathcal{P}(x,D)u \in \mathcal{S}(\mathbb{R}^n)$ if $u \in \mathcal{S}(\mathbb{R}^n)$. We consider the expression $(1+|x|^2)^p \partial_x^\alpha(\mathcal{P}(x,D)u)$ which equals

$$(1+|x|^2)^p \partial_x^\alpha \left((2\pi)^{-n/2} \int_{\mathbb{R}^n} e^{i<x,\xi>} p(x,\xi)\hat{u}(\xi)d\xi \right) =$$
$$(1+|x|^2)^p (2\pi)^{-n/2} \int_{\mathbb{R}^n} \hat{u}(\xi) \left[\sum_{0\leq\beta\leq\alpha} \binom{\alpha}{\beta} \partial_x^\beta p(x,\xi)(i\xi)^{\alpha-\beta} e^{i<x,\xi>} \right] d\xi \qquad (1.18)$$

A typical term in (1.18) appears to be

$$\int_{\mathbb{R}^n} (1+|x|^2)^p \partial_x^\beta p(x,\xi)(i\xi)^{\alpha-\beta} e^{i<x,\xi>} \hat{u}(\xi)d\xi \qquad (1.19)$$

We next use (1.16) – with $\alpha = (0,\ldots 0)$, so that $(1+|x|^2)^p|\partial_x^\beta p(x,\xi)| \leq C_{p,\beta}(1+|\xi|)^r$. Then, the term under integral sign in (1.19) is estimated in absolute value by

$$C_{p,\beta}(1+|\xi|)^r |\xi|^{|\alpha-\beta|} |\hat{u}(\xi)| \leq C_{1,p}(1+|\xi|^2)^{r/2}(1+|\xi|^2)^{|\alpha-\beta|/2-p},$$
$$\forall p = 1,2,\ldots \qquad (1.20)$$

(we justify (1.20) in the following way: first, $(1+x)^r \leq C(1+x^2)^{r/2}$ holds, $\forall x \geq 0$, $\forall r \in \mathbb{R}$; in fact, we see that $(1+x^2) \leq (1+x)^2 \leq 2(1+x^2) \forall x \geq 0$; if $r > 0$ we get $(1+x)^r \leq 2^{r/2}(1+x^2)^{r/2}$; if $r < 0$ we derive $(1+x^2)^{r/2} \geq (1+x)^r$.

Next $(i\xi)^{\alpha-\beta}$ means $(\sqrt{-1})^{|\alpha-\beta|}\xi_1^{\alpha_1-\beta_1}\xi_2^{\alpha_2-\beta_2}\ldots\xi_n^{\alpha_n-\beta_n}$, where $|\alpha-\beta| = \sum_1^n(\alpha_i-\beta_i)$.

As we know the obvious estimates: $|\xi_j| \leq |\xi|\ \forall j = 1,2,\ldots,n$, we obtain $|(i\xi)^{\alpha-\beta}| = |\xi_1|^{\alpha_1-\beta_1}\ldots|\xi_n|^{\alpha_n-\beta_n} \leq |\xi|^{|\alpha-\beta|} \leq (1+|\xi|)^{|\alpha-\beta|} \leq C(1+|\xi|^2)^{|\alpha-\beta|/2}$; finally, as $\hat{u}(\xi) \in \mathcal{S}$, we have also $|\hat{u}(\xi)| \leq \Gamma_p(1+|\xi|^2)^{-p}$, $\forall p \in \underline{\mathbb{N}}$, $\forall \xi \in \mathbb{R}^n$).

Note now that the right-hand side in (1.20) belongs to $L^1(\mathbb{R}^n)$ if p is sufficiently large; therefore we obtain from (1.18) that

$$\left|(1+|x|^2)^p \partial_x^\alpha(\mathcal{P}(x,D)u)\right| \leq \Gamma_p,\ \forall x \in \mathbb{R}^n,\ \forall \alpha \in \underline{\mathbb{N}}^n \tag{1.21}$$

which shows in fact that $\mathcal{P}(x,D)u \in \mathcal{S}(\mathbb{R}^n)$.

Let us use now Th. 10.1 in [9] – p.97; we find the estimates

$$\| \mathcal{P}(x,D)u \|_{H^s} \leq C_s \| u \|_{H^{s+r}},\ \forall s \in \mathbb{R},\ \forall u \in \mathcal{S}(\mathbb{R}^n) \tag{1.22}$$

Accordingly, taking $E^\infty = H^\infty$ and $V = \mathcal{S}(\mathbb{R}^n)$, we infer from (1.22) that r belongs to the order set of $\mathcal{P}(x,D)$.

2. Two subalgebras of $\mathcal{L}in(V)$

We shall consider here the subalgebras of $\mathcal{L}in(V)$ (over the field K), denoted respectively with $\mathcal{L}in_{\mathcal{O}\neq\phi}(V)$ and $\mathcal{L}in_{\mathcal{O}=\mathbb{R}}(V)$.

The first one consists of all $L \in \mathcal{L}in(V)$ which have a *non-empty* order set; the second one consist of those $L \in \mathcal{L}in(V)$ having order set = the real line.

Proposition 2.1 *The set $\mathcal{L}in_{\mathcal{O}\neq\phi}(V)$ is a subalgebra of $\mathcal{L}in(V)$ over K*

To establish this simple result, let us take L_1, L_2 in $\mathcal{L}in_{\mathcal{O}\neq\phi}(V)$. If $r_1 \in \mathcal{O}(L_1)$ and $r_2 \in \mathcal{O}(L_2)$ we have $\| L_i u \|_{E^s} \leq C_{s,r_i} \| u \|_{E^{s+r_i}}$, $\forall u \in V$, where $i = 1,2$.

Consequently we get $\| (L_1 + L_2)u \|_{E^s} \leq (C_{s,r_1} + C_{s,r_2}) \| u \|_{E^{s+r}}$ where $r = \max(r_1, r_2)$. Thus, $r \in \mathcal{O}(L_1 + L_2)$.

Next, let $\lambda \in K$ and $r \in \mathcal{O}(L)$; then $\| Lu \|_{E^s} \leq C_{s,r} \|_{E^{s+r}}, \forall u \in V$ and $\| \lambda L u \|_{E^s} = |\lambda| \| Lu \|_{E^s} \leq |\lambda| C_{s,r} \| u \|_{E^{s+r}}, \forall u \in V$. Therefore $r \in \mathcal{O}(\lambda L)$.

Finally, again with L_1, L_2 in $\mathcal{L}in_{\mathcal{O}\neq\phi}(V), r_1 \in \mathcal{O}(L_1), r_2 \in \mathcal{O}(L_2)$ we have

$$\| (L_1 \cdot L_2)u \|_{E^s} \leq C^1_{s,r_1} \| L_2 u \|_{E^{s+r_1}} \leq C^1_{s,r_1} . C^2_{s,r_2} \| u \|_{E^{s+r_1+r_2}} \quad \forall u \in V \tag{2.1}$$

This shows that $r_1 + r_2 \in \mathcal{O}(L_1 \cdot L_2)$.

In the same way we see that $r_1 + r_2 \in \mathcal{O}(L_2 \cdot L_1)$.

∎

Remark In any case, the number $r = 0$ belongs to the order set of the identity mapping, $V \to V$. Therefore $\mathcal{L}in_{\mathcal{O}\neq\phi}(V)$ is a subalgebra of $\mathcal{L}in(V)$ with unit element I (identity).

Proposition 2.2 *The set $\mathcal{L}in_{\mathcal{O}=\mathbb{R}}(V)$ is also a subalgebra of $\mathcal{L}in(V)$*

In fact, take any $r \in \mathbb{R}$. If $L_1, L_2 \in \mathcal{L}in_{\mathcal{O}=\mathbb{R}}(V)$, we have

$$\| L_i\nu \|_{E^s} \leq C_{s,r} \| \nu \|_{E^{s+r}}, \ \forall \nu \in V, \ \forall s \in \mathbb{R}. \text{ (where } i = 1.2) \tag{2.2}$$

Then $r \in \mathcal{O}(L_1 + L_2)$ (see Prop. 2.1).

Also we have:

$$\| (L_1 \cdot L_2)u \|_{E^s} \leq C^1_{s,r/2} \| L_2 \|_{E^{s+r/2}} \leq C^1_{s,r/2} \cdot C^2_{s,r/2} \| u \|_{E^{s+r}} \tag{2.3}$$

$\forall u \in V$; thus $r \in \mathcal{O}(L_1 \cdot L_2)$.

In a similar way, we see that λL has order set $= \mathbb{R}$ (for $\lambda \in \mathbb{C}$) and that $\mathcal{O}(L_2 \cdot L_1) = \mathbb{R}$ too.

Remark 2.1 If $\mathcal{O}(L) = \mathbb{R}$, the "true order" of L defined by $t.oL = \inf \mathcal{O}(L)$ will be $-\infty$. This justifies the notation $\mathcal{L}_{-\infty} = \mathcal{L}in_{\mathcal{O}=\mathbb{R}}(V)$. On the other hand operators L with $\mathcal{O}(L) = \phi$ have true order $= +\infty$ (by definition!).

Remark 2.2 *If the identity mapping in V belongs to $\mathcal{L}_{-\infty}$, then all the norms $\| \ \|_{E^s}$ are equivalent on V.*

In fact we must have inequalities

$$\| Iu \|_{E^s} = \| u \|_{E^s} \leq C_{s,-p} \| u \|_{E^{s-p}}, \ \forall s \in \mathbb{R}, \ \forall p \in \mathbb{R}, \ \forall u \in V \tag{2.4}$$

If $p > 0$ we get (using also (1.2))

$$\| u \|_{E^{s-p}} \leq \| u \|_{E^s} \leq C_{s,-p} \| u \|_{E^{s-p}}, \ \forall s \in \mathbb{R}, \ \forall u \in V. \tag{2.5}$$

∎

Remark 2.3 *$\mathcal{L}_{-\infty}$ is a right and left ideal in $\mathcal{L}in_{\mathcal{O}\neq\phi}(V)$.*

This assertion means the following:

if $T \in \mathcal{L}_{-\infty}$ and $L \in \mathcal{L}in_{\mathcal{O}\neq\phi}(V)$, then $T \cdot L$ and $L \cdot T$ belong to $\mathcal{L}_{-\infty}$.

Proof In fact, in Prop. 2.1 we have seen that: $r_1 \in \mathcal{O}(L_1)$, $r_2 \in \mathcal{O}(L_2) \Rightarrow r_1 + r_2 \in \mathcal{O}(L_1 \cdot L_2)$, $r_1 + r_2 \in \mathcal{O}(L_2 \cdot L_1)$.

Hence, if $r \in \mathcal{O}(L)$ and if $s \in \mathbb{R}$, we take $\sigma \in \mathcal{O}(T)$ such that $r + \sigma = s$. Thus $s \in \mathcal{O}(T \cdot L)\ \forall s \in \mathbb{R}$, and $s \in \mathcal{O}(L \cdot T)\ \forall s \in \mathbb{R}$; it follows that $\mathcal{O}(T \cdot L) = \mathcal{O}(L \cdot T) = \mathbb{R}$, $T \cdot L \in \mathcal{L}_{-\infty}, L \cdot T \in \mathcal{L}_{-\infty}$.

3. Structure of order sets

Our main goal in this section : to prove the following result:

Theorem 3.1 *Let $(E^s)_{s \in \mathbb{R}}$ be a scale of Banach spaces over the same field ($\mathbb{R}$ or $\mathbb{C}$); let $E^\infty = \bigcap_{s \in \mathbb{R}} E^s$ and V be a vector subspace of E^∞. Let $L \in \mathcal{L}in(V)$. Then the order set $\mathcal{O}(L)$ must be one of the four kinds below:*

i) the empty set ϕ

ii) the real line $\mathbb{R}$

iii) an open half-interval (a, ∞), where $a \in \mathbb{R}$

iv) a closed half-interval $[a, \infty)$, where $a \in \mathbb{R}$.

Proof i) Let E be a Banach space and then $\{E^s\}_{s \in \mathbb{R}}$ be a "constant" scale: $E^s = E \forall s \in \mathbb{R}$. Then $E^\infty = E$ and we also take $V = E^\infty = E$.

Next, consider a linear operator $L, E \to E$. If L *is not* continuous, an estimate $\| Lu \|_{E^s} = \| Lu \|_E \leq C \| u \|_{E^s + r} = C \| u \|_E \quad \forall u \in E$ is impossible. Therefore $\mathcal{O}(L) = \phi$.

ii) Take again the same (constant) scale as above; then, looking at any linear *continuous* $L, E \to E$, we see that *any* real number r belongs to $\mathcal{O}(L)$.

(Another example appeared in previous discussions: we took $V = H^\infty$, and $L = \varphi(D) = \mathcal{F}^{-1} \mathcal{M}_{\varphi(\cdot)} \mathcal{F}$, where $\varphi(\xi)$ is a continuous function $\mathbb{R}^n \to \mathbb{C}$, vanishing outside some compact subset of $\mathbb{R}^n$. It has been seen that $\mathcal{O}(\varphi(D)) = \mathbb{R}$).

iii) In general, $\mathcal{O}(L)$ is a subset of $\mathbb{R}$. Assume it is not bounded from below. Then, $\mathcal{O}(L) = \mathbb{R}$; in fact $\forall r \in \mathbb{R}, \exists r' < r$ and $r' \in \mathcal{O}(L)$. On the other hand, we have seen that if $a \in \mathcal{O}(L)$ and $a' > a \Rightarrow a' \in \mathcal{O}(L)$ too. Accordingly, $r \in \mathcal{O}(L)$. This is the situation covered in (ii) – when $\mathcal{O}(L) = \mathbb{R}$. Assume now that $\mathcal{O}(L)$ *is* lower bounded and let $\ell = \inf \mathcal{O}(L) =$ G.L.B. $\mathcal{O}(L)$. Then, the open interval $(\ell, +\infty)$ is contained in $\mathcal{O}(L)$; for, if $\ell_1 > \ell$, there exists $r \in \mathcal{O}(L), \ell < r \leq \ell_1$; this implies that $\ell_1 \in \mathcal{O}(L)$.

We shall now see an example where $\ell \notin \mathcal{O}(L)$.

Take

$$E^s = \mathcal{B}_{1,s}(\mathbb{R}^n), V = E^\infty = \bigcap_{s\in\mathbb{R}} \mathcal{B}_{1,s}(\mathbb{R}^n). \tag{3.1}$$

Consider the function $\mathbb{R}^n \to \mathbb{R}^+$, given by: $\xi \in \mathbb{R}^n \to \ln(2+|\xi|)$. Remember the elementary estimate: $\forall\varepsilon > 0, \exists c_\varepsilon > 0$, such that $\ln(2+|\xi|) \leq c_\varepsilon(1+|\xi|^2)^{\varepsilon/2}$, $\forall\xi \in \mathbb{R}^n$. It will follow that the "Friedrichs" operator: $\mathcal{F}^{-1}\mathcal{M}_{\varphi(\cdot)}\mathcal{F}$ associated to $\varphi(\xi) = \ln(2+|\xi|)$ is a well-defined (linear) mapping from E^∞ into itself: in fact, if $u \in E^\infty$ we have that $(1+|\xi|^2)^{s/2}\hat{u}(\xi) \in L^1(\mathbb{R}^n)$ $\forall s \in \mathbb{R}$. Consequently: $|\ln(2+|\xi|)\hat{u}(\xi)| \leq c_\varepsilon(1+|\xi|^2)^{\varepsilon/2}|\hat{u}(\xi)|$, $\forall\xi \in \mathbb{R}^n$ and $(1+|\xi|^2)^{s/2}|\ln(2+|\xi|)\hat{u}(\xi)| \leq c_\varepsilon(1+|\xi|^2)^{(s+\varepsilon)/2}|\hat{u}(\xi)|$, $\forall\xi \in \mathbb{R}^n$. This shows that $(1+|\xi|^2)^{s/2}|\ln(2+|\xi|)\hat{u}(\xi)| \in L^1(\mathbb{R}^n)$ $\forall s \in \mathbb{R}$. Thus, $\forall u \in E^\infty, \ln(2+|D|)u$ belongs to E^∞ and, furthermore, we have

$$\| \ln(2+|D|)u \|_{E^s}$$

$$= \int_{\mathbb{R}^n} (1+|\xi|^2)^{s/2}|\ln(2+|\xi|)\hat{u}(\xi)|d\xi \leq c_\varepsilon \| u \|_{E^{s+\varepsilon}}, \ \forall u \in E^\infty, \ \forall s \in \mathbb{R}, \ \forall\varepsilon > 0.$$

Therefore, any $\varepsilon > 0$ belongs to $\mathcal{O}(\ln(2+|D|))$. However $r = 0$ *itself does not belong to this order set.* This is shown as follows: assume $0 \in \mathcal{O}(\ln(2+|D|))$. Then, the estimates

$$\| \ln(2+|D|)u \|_{E^s} \leq C_s \| u \|_{E^s}, \ \forall s \in \mathbb{R}, \ \forall u \in E^\infty \tag{3.2}$$

are verified.

In particular, they are verified for $s = 0$, that is in the space $\mathcal{B}_{1,0}(\mathbb{R}^n) = \mathcal{F}^{-1}(L^1(\mathbb{R}^n))$. It follows that

$$\int_{\mathbb{R}^n} \ln(2+|\xi|)|\hat{u}(\xi)|d\xi \leq c\int_{\mathbb{R}^n} |\hat{u}(\xi)|d\xi, \ \forall u \in \mathcal{B}_{1,\infty}(\mathbb{R}^n) \tag{3.2}$$

We know that $\mathcal{S}(\mathbb{R}^n) \subset \mathcal{B}_{1,\infty}(\mathbb{R}^n)$, and we shall apply (3.2) to a specially chosen sequence $(u_p)_1^\infty$ in $\mathcal{S}(\mathbb{R}^n)$. Precisely, let us first take functions $f_p(t), [0,\infty) \to \mathbb{R}, f_p(\cdot) \in C^\infty, f_p(t) = 0$ for $0 \leq t \leq p$ and for $4p \leq t < +\infty; f_p(t) = \frac{1}{p}$ for $2p \leq t \leq 3p; 0 \leq f_p(t) \leq \frac{1}{p}$ for all $t \geq 0$; then define functions $g_p(\xi), \mathbb{R}^n \to \mathbb{R}$, by the relation: $g_p(\xi) = f_p(|\xi|)$. We see that $g_p(\xi) = 0$ for $|\xi| \leq p$ and for $|\xi| \geq 4p; g_p(\xi) = \frac{1}{p}$ for $2p \leq |\xi| \leq 3p; 0 \leq g_p(\xi) \leq \frac{1}{p}$ for all $\xi \in \mathbb{R}^n; g_p(\xi) \in C^\infty(\mathbb{R}^n); g_p(\xi)$ have compact support in $\mathbb{R}^n$ (which is contained in the ball $\{\xi \in \mathbb{R}^n, |\xi| \leq 4p\}$.)

Accordingly, the sequence of functions: $u_p(x) = \mathcal{F}^{-1}(g_p(\xi))$ is composed of $\mathcal{S}(\mathbb{R}^n)$-functions, $\forall p = 1, 2, \ldots$, and $\hat{u}_p(\xi) = g_p(\xi)$.

Let us write now (3.2) for this sequence $(u_p(\cdot))$. Note that

$$\int_{\mathbb{R}^n} \ln(2+|\xi|)|\hat{u}_p(\xi)|d\xi \geq \int_{2p\leq|\xi|\leq 3p} g_p(\xi)\ln(2+|\xi|)d\xi = \frac{1}{p}\int_{2p\leq|\xi|\leq 3p} \ln(2+|\xi|)d\xi \geq$$

$$\frac{1}{p}\ln(2+2p)m(\xi; 2p\leq|\xi|\leq 3p) = c\frac{1}{p}\ln(2+2p)((3p)^n - (2p)^n), \tag{3.3}$$

while

$$\int_{\mathbb{R}^n} |\hat{u}_p(\xi)|d\xi = \int_{\mathbb{R}^n} |g_p(\xi)|d\xi = \int_{p\leq|\xi|\leq 4p} |g_p(\xi)|d\xi \leq \frac{1}{p}\cdot c((4p)^n - p^n). \tag{3.4}$$

Therefore, using (3.2), we would get:

$$\frac{c}{p}\ln(2+2p)((3p)^n - (2p)^n) \leq \frac{c}{p}((4p)^n - p^n) \tag{3.5}$$

that is, after simplification;

$$\ln(2+2p) \leq \frac{4^n - 1}{3^n - 2^n}, \quad \forall p = 1, 2, \ldots \tag{3.6}$$

which, of course, is impossible. (Note therefore that $t.o(\ln(2+|D|)) = 0$).

iv) In this (final) part of our proof we shall give an example of an operator L whose order set is the *closed* half interval $[0, +\infty)$ (and therefore, $t.o(L) = 0$ again). We take $E^s = H^s(\mathbb{R})$ and $V = H^\infty(\mathbb{R})$. Let $\psi(\xi) = 1$ for $\xi \geq 0, \psi(\xi) = 0$ for $\xi < 0$, and then, as usual, $\psi(D) = L = \mathcal{F}^{-1}\mathcal{M}_{\psi(\cdot)}\mathcal{F}$. Using for instance estimates (1.12), we get: $\| \psi(D)u \|_{H^s} \leq \| u \|_{H^s}$, $\forall u \in H^\infty$, $\forall s \in \mathbb{R}$. Therefore $[0, \infty) \subset \mathcal{O}(\psi(D))$; thus, in order to establish the equality $\mathcal{O}(\psi(D)) = [0, \infty)$ it will suffice to show that no negative real number can belong to $\mathcal{O}(\psi(D))$; this amounts to demonstrate that an inequality

$$\| \psi(D)u \|_{H^s} \leq C_{s,\varepsilon} \| u \|_{H^{s-\varepsilon}}, \; \forall s \in \mathbb{R}, \; \forall u \in H^\infty \tag{3.7}$$

is impossible when $\varepsilon > 0$.

In fact, assuming the contrary (that is (3.7) – holds true) we obtain, taking the special case where $s = 0$, that

$$\| \psi(D)u \|_{L^2(\mathbb{R})} \leq C_\varepsilon \| u \|_{H^{-\varepsilon}(\mathbb{R})}, \; \forall u \in \mathcal{S}(\mathbb{R}) \tag{3.8}$$

(for some $\varepsilon > 0$).

Consider now a sequence $g_p(\xi) \in C_0^\infty(\mathbb{R})$, where $0 \leq g_p(\xi) \leq 1, g_p(\xi) = 0$ for $\xi \leq p-1$ and for $\xi \geq 2p+1, g_p(\xi) = 1$ for $p \leq \xi \leq 2p$; next, let $u_p(x) = \mathcal{F}^{-1}(g_p(\cdot)), p = 1, 2, \ldots$ Thus $u_p(\cdot) \in \mathcal{S}(\mathbb{R})$ $\forall p \in \mathbb{N}$. Now, let us apply (3.8) to $u = u_p$; it will mean the sequence of estimates

$$\int_{\mathbb{R}} |\psi(\xi)\hat{u}_p(\xi)|^2 d\xi \leq c_\varepsilon^2 \int_{\mathbb{R}} (1+|\xi|^2)^{-\varepsilon} |\hat{u}_p(\xi)|^2 d\xi, \ \forall p = 1, 2, \ldots \tag{3.9}$$

On the other hand we note that

$$\int_{\mathbb{R}} |\psi(\xi)\hat{u}_p(\xi)|^2 d\xi = \int_0^\infty |\hat{u}_p(\xi)|^2 d\xi = \int_0^\infty |g_p(\xi)|^2 d\xi \geq \int_p^{2p} d\xi = p \tag{3.10}$$

and also that

$$\int_{\mathbb{R}} (1+|\xi|^2)^{-\varepsilon} |g_p(\xi)|^2 d\xi \leq \int_{p-1}^{2p+1} (1+|\xi|^2)^{-\varepsilon} d\xi \leq (p+2)(1+(p-1)^2)^{-\varepsilon}, \ \forall p \in \mathbb{N}. \tag{3.11}$$

Hence, using (3.9)–(3.10)–(3.11), we find that $p \leq c_\varepsilon^2 (p+2)(1+(p-1)^2)^{-\varepsilon}$, $\forall p \in \mathbb{N}$, that is

$$\frac{p}{p+2} \leq \frac{c_\varepsilon^2}{(1+(p-1)^2)^\varepsilon}, \ \forall p \in \mathbb{N} \tag{3.12}$$

which is impossible (just let $p \to \infty$!).

This terminates proof of Theorem 3.1.

Remark In view of above result we can see that, $\forall L_1, L_2$ in $\mathcal{L}in(V)$ we have $\mathcal{O}(L_1) \subseteq \mathcal{O}(L_2)$ or $\mathcal{O}(L_2) \subset \mathcal{O}(L_1)$.

4. Properties of the order sets and of the true order

First we remember that if A, B are subsets of $\mathbb{R}$, their vector sum $A+B$ is defined as $A+B = \{a+b, a \in A, b \in B\}$. However, if $A = \phi$ or $B = \phi$ we put $A+B = \phi$.

We shall first establish

Proposition 4.1 *If $L_1, L_2 \in \mathcal{L}in(V)$, we have:*

a) $\mathcal{O}(L_1) + \mathcal{O}(L_2) \subset \mathcal{O}(L_1 L_2)$ and $\mathcal{O}(L_1) + \mathcal{O}(L_2) \subset \mathcal{O}(L_2 L_1)$

b) the equality $\mathcal{O}(L_1) + \mathcal{O}(L_2) = \mathcal{O}(L_1 L_2)$ is, in general, false.

Proof a) If $\mathcal{O}(L_1) = \phi$ or $\mathcal{O}(L_2) = \phi$, we have $\mathcal{O}(L_1) + \mathcal{O}(L_2) = \phi \subset \mathcal{O}(L_1 L_2)$. If $\mathcal{O}(L_1)$ and $\mathcal{O}(L_2)$ are nonempty we apply a previous result (see proof of Prop. 2.1): if $r_1 \in \mathcal{O}(L_1)$ and $r_2 \in \mathcal{O}(L_2)$ then $r_1 + r_2 \in \mathcal{O}(L_1 . L_2) \cap \mathcal{O}(L_2 . L_1)$.

b) Take $E^s = H^s(\mathbb{R}), V = H^\infty(\mathbb{R})$. Consider two functions: $\psi_1(\xi) = 1$ for $\xi \geq 0$ and $= 0$ for $\xi < 0, \psi_2(\xi) = 0$ for $\xi \geq 0, = 1$ for $\xi < 0$. As seen in Th. 3.1 (iv), the operator $\psi_1(D)$ has the order set $[0, \infty)$. With very much the same proof we can see that the operator $\psi_2(D)$ has again $[0, \infty)$ as order set. Then $\mathcal{O}(\psi_1(D)) + \mathcal{O}(\psi_2(D)) = [0, \infty) + [0, \infty) = [0, \infty)$. However, the product operator $L_1 . L_2$ is defined by (for $u \in H^\infty$): $[\psi_1(D)(\psi_2(D)u)]\hat{}(\xi) = \psi_1(\xi)(\psi_2(\xi)\hat{u}(\xi)) = 0\ \forall \xi \in \mathbb{R}$. Hence $\psi_1(D) . \psi_2(D)$ is the null operator in $H^\infty(\mathbb{R})$ which has, obviously, order set $= \mathbb{R}$.

There is a corresponding result for the *true order*:

Proposition 4.2 *Assume $L_1, L_2 \in \mathcal{L}in(V)$ and $\mathcal{O}(L_1) \neq \phi, \mathcal{O}(L_2) \neq \phi$. Then $t.o\ (L_1 . L_2) \leq t.o(L_1) + t.o(L_2)$ while the equality: $t.o(L_1 . L_2) = t.o(L_1) + t.o(L_2)$ is, in general, false.*

Proof The last assertion follows from part b) of previous proof.

The first assertion is established as follows:

For any $\varepsilon > 0, \exists r_1 \in \mathcal{O}(L_1)$ such that $t.o(L_1) \leq r_1 < t.o(L_1) + \varepsilon$ and $\exists r_2 \in \mathcal{O}(L_2)$ such that $t.o(L_2) \leq r_2 < t.o(L_2) + \varepsilon$. Hence:

$t.o(L_1) + t.o(L_2) \leq r_1 + r_2 < t.o(L_1) + t.o(L_2) + 2\varepsilon$.

On the other hand, we know that $r_1 + r_2 \in \mathcal{O}(L_1 . L_2)$. Hence $t.o(L_1 . L_2) \leq r_1 + r_2 < t.o(L_1) + t.o(L_2) + 2\varepsilon$, $\forall \varepsilon > 0$. This gives the result when $-\infty < t.o(L_1)$ and $-\infty < t.o(L_2)$.

In the case when, say, $t.o(L_1) = -\infty$, we have $\mathcal{O}(L_1) = \mathbb{R}$. Hence $\mathcal{O}(L_1) + \mathcal{O}(L_2) = \mathbb{R}$ too, and by Prop. 4.1, $\mathcal{O}(L_1 L_2) = \mathbb{R}$. We have now the "equality":

$$t.o(L_1 L_2) = -\infty = t.o(L_1) + t.o(L_2)$$

(note that, from assumption, $t.o(L_1) < +\infty, t.o(L_2) < +\infty$).

Next we shall establish

Proposition 4.3 *If $\lambda \in K/\{0\}$ and $L \in \mathcal{L}in(V)$, then $\mathcal{O}(\lambda L) = \mathcal{O}(L)$.*

Proof i) We first consider the situation where $\mathcal{O}(L) = \phi$. We must prove that, for $\lambda \neq 0, \mathcal{O}(\lambda L) = \phi$. If not, $\mathcal{O}(\lambda L) \neq \phi$ and $\exists r \in \mathbb{R}, r \in \mathcal{O}(\lambda L)$. It follows that, $\forall s \in \mathbb{R}$

$\| (\lambda L)u \|_{E^s} \leq C_{s,r} \| u \|_{E^{s+r}}$, $\forall u \in V$; but this is turn gives, again $\forall s \in \mathbb{R}$, the estimate

$$\| Lu \|_{E^s} = \| \frac{1}{\lambda}(\lambda L)u \|_{E^s} = \frac{1}{|\lambda|} \| (\lambda L)u \|_{E^s} \leq \frac{1}{|\lambda|} C_{s,r} \| u \|_{E^{s+r}}, \ \forall u \in V$$

Consequently $r \in \mathcal{O}(L)$, and $\mathcal{O}(L) \neq \phi$, a contradiction.

ii) Here we first remember (see proof of Prop. 2.1), that, if $\mathcal{O}(L) \neq \phi$ then $\forall \lambda \in K$, $\mathcal{O}(L) \subset \mathcal{O}(\lambda L)$ (this inclusion can be strict; if $\lambda = 0$, we take L with $\mathcal{O}(L) \neq \mathbb{R}$, and we note that $\mathcal{O}(\Theta) = \mathbb{R}$).

Take therefore again $\lambda \neq 0$. We obtain:

$\mathcal{O}(\lambda L) \subset \mathcal{O}(\frac{1}{\lambda}(\lambda L)) = \mathcal{O}(L)$; hence the equality: $\mathcal{O}(L) = \mathcal{O}(\lambda L)$

(for instance we always have: $\mathcal{O}(-L) = \mathcal{O}(L)$).

Let us now say something about $\mathcal{O}(L_1 + L_2)$, where $L_1, L_2 \in \mathcal{L}in(V)$.

Proposition 4.4 *The equalities*

$\mathcal{O}(L_1 + L_2) = \mathcal{O}(L_1) + \mathcal{O}(L_2)$ *(as vector sum) and*

$\mathcal{O}(L_1 + L_2) = \mathcal{O}(L_1) \cup \mathcal{O}(L_2)$ *(as set-theoretic union)*

are, in general, false.

Proof Take $L_1 = L, L_2 = -L$ where $\mathcal{O}(L) = \mathcal{O}(-L) = [a, +\infty)$. Then $\mathcal{O}(L_1 + L_2) = \mathcal{O}(\Theta) = \mathbb{R}$, while $[a, \infty) + [a, \infty) = [2a, +\infty)$ and $[a, \infty) \cup [a, \infty) = [a, \infty)$.

Next, we note the

Proposition 4.5 *For any* $L_1, L_2 \in \mathcal{L}in(V)$ *we have*

$\mathcal{O}(L_1 + L_2) \supset \mathcal{O}(L_1)$ *if* $\mathcal{O}(L_1) \subseteq \mathcal{O}(L_2)$ *or*

$\mathcal{O}(L_1 + L_2) \supset \mathcal{O}(L_2)$ *if* $\mathcal{O}(L_2) \subseteq \mathcal{O}(L_1)$

Proof The result is obvious if $\mathcal{O}(L_1)$ or $\mathcal{O}(L_2)$ is the empty set.

Otherwise, let $\mathcal{O}(L_1) \subset \mathcal{O}(L_2)$, and take $r \in \mathcal{O}(L_1)$. It results $\| L_1 u \|_{E^s} \leq C_{1,s} \| u \|_{E^{s+r}}$ and $\| L_2 u \|_{E^s} \leq C_{2,s} \| u \|_{E^{s+r}}$, $\forall s \in \mathbb{R}$, $\forall u \in V$.

Therefore we obtain

$\| (L_1 + L_2)u \|_{E^s} \leq (C_{1,s} + C_{2,s}) \| u \|_{E^{s+r}}$, $\forall s \in \mathbb{R}$, $\forall u \in V$, whence $r \in \mathcal{O}(L_1 + L_2)$.

■

Remark The inclusions above can be strict: take $\mathcal{O}(L) = [0, \infty) = \mathcal{O}(-L)$ and note that $\mathcal{O}(\Theta) = \mathbb{R}$.

We may also inquire about validity of inclusions:

$\mathcal{O}(L_1+L_2) \supset \mathcal{O}(L_1)+\mathcal{O}(L_2)$ or $\mathcal{O}(L_1+L_2) \supset \mathcal{O}(L_1) \cup \mathcal{O}(L_2)$

They are both false, in general :

i) Take $E^s = E \; \forall s \in \mathbb{R}$, where E is a fixed Banach space. Then take L_1 – a linear continuous operator in E, and L_2 – a linear discontinuous operator $E \to E$ and we have accordingly: (as $L_1 + L_2$ is also discontinuous)

$\mathcal{O}(L_1) = \mathbb{R}, \mathcal{O}(L_2) = \phi, \mathcal{O}(L_1+L_2) = \phi$. Then $\mathcal{O}(L_1) \cup \mathcal{O}(L_2) = \mathbb{R} \cup \phi = \mathbb{R}$, and $\mathbb{R} \not\subset \phi$.

ii) Take $V = E^\infty = \bigcap_{s\in\mathbb{R}} \mathcal{B}_{1,s}(\mathbb{R}^n), L_1 = \Theta, L_2 = \ln(2+|D|)$ (see Th. 3.1 – (iii)). We obtain $L_1+L_2 = \ln(2+|D|), \mathcal{O}(L_1+L_2) = (0,\infty), \mathcal{O}(L_1) = \mathbb{R}, \mathcal{O}(L_2) = (0,\infty)$, and the inclusion: $(0,\infty) \supset \mathbb{R} + (0,\infty) = \mathbb{R}$ is false.

To end this section we note

Proposition 4.6 *For any $L_1, L_2 \in \mathcal{L}in(V)$ we have*

$$t.o(L_1+L_2) \le \max(t.o(L_1), t.o(L_2)).$$

Proof This is a Corollary to Prop. 4.5. We get:

$t.o(L_1) \ge t.o(L_1+L_2)$ if $\mathcal{O}(L_1) \subseteq \mathcal{O}(L_2)$ or

$t.o(L_2) \ge t.o(L_1+L_2)$ if $\mathcal{O}(L_2) \subseteq \mathcal{O}(L_1)$

On the other hand, $\mathcal{O}(L_1) \subseteq \mathcal{O}(L_2), \Rightarrow t.o(L_2) \le t.o(L_1)$, while $\mathcal{O}(L_2) \subseteq \mathcal{O}(L_1) \Rightarrow t.o(L_1) \le t.o(L_2)$.

5. Some more concrete examples

In this section we give (again) some explicit "computation" of the true order for some operators in $H^\infty(\mathbb{R}^n) = \bigcap_{s\in\mathbb{R}} H^s(\mathbb{R}^n) = V$.

The first result is similar to the Example in Th. 3.1 (iv).

Proposition 5.1 *Let $\psi(\xi)$ be a C^∞-function, $\mathbb{R}^n \to \mathbb{C}$, such that $\psi(\xi) = 0$ for $|\xi| \le \frac{1}{2}, \psi(\xi) = 1$ for $|\xi| \ge 1, 0 \le \psi(\xi) \le 1 \; \forall \xi \in \mathbb{R}^n$.*

Consider the operator $\psi(D) = \mathcal{F}^{-1}\mathcal{M}_{\psi(\cdot)}\mathcal{F}, H^\infty \to H^\infty$. Then we have $t.o(\psi(D)) = 0$, and the order set of $\psi(D)$ is the interval $[0,\infty)$.

Proof The function $\psi(\cdot)$ here is bounded and measurable on $\mathbb{R}^n$; hence, as seen in (1.11)–(1.12), $[0,\infty) \subset \mathcal{O}(\psi(D))$. Thus, we only have to establish that $\nexists r < 0$

and $r \in \mathcal{O}(\psi(D))$. Suppose instead that such a number r exists. It will follow then

$$\| \psi(D)u \|_{H^s} \leq C \| u \|_{H^{s+r}}, \ \forall u \in \mathcal{S}(\mathbb{R}^n), \ \forall s \in \mathbb{R}. \tag{5.1}$$

In particular, if $s = 0$, we would have the estimate

$$\int_{\mathbb{R}^n} |\psi(\xi)|^2 |\hat{u}(\xi)|^s d\xi \leq C \int_{\mathbb{R}^n} (1 + |\xi|^2)^r |\hat{u}(\xi)|^2 d\xi, \ \forall u \in \mathcal{S}(\mathbb{R}^n) \tag{5.2}$$

therefore

$$\int_{|\xi|\geq 1} |\hat{u}(\xi)|^2 d\xi \leq C \int_{\mathbb{R}^n} (1 + |\xi|^2)^r |\hat{u}(\xi)|^2 d\xi, \ \forall u \in \mathcal{S}(\mathbb{R}^n). \tag{5.3}$$

We shall apply this estimate to a special sequence of functions $u_p(\cdot) \in \mathcal{S}(\mathbb{R}^n)(p \in \mathbb{N})$:

Precisely, let $f_p(\xi) \in C_0^\infty(\mathbb{R}^n), f_p(\xi) = 0$ for $|\xi| \geq 2p$, and $f_p(\xi) = 1$ for $|\xi| \leq p$, $0 \leq f_p(\xi) \leq 1$ (put $f_p(\xi) = f(\frac{\xi}{p})$ where $f \in C_0^\infty(\mathbb{R}^n), f = 0$ for $|\xi| \geq 2, f = 1$ for $|\xi| \leq 1, 0 \leq f \leq 1$).

Next, take $g(\xi) = (1 + |\xi|^2)^{-n/4}, \xi \in \mathbb{R}^n$; thus $g(\cdot) \in C^\infty(\mathbb{R}^n)$. Define the sequence: $\varphi_p(\xi) = f_p(\xi)g(\xi)$; consequently, $\varphi_p(\xi) = 0$ for $|\xi| \geq 2p, \varphi_p(\xi) = (1 + |\xi|^2)^{-n/4}$ for $|\xi| \leq p, \varphi_p(\cdot) \in C_0^\infty(\mathbb{R}^n)$. Finally, we put: $u_p(x) = \mathcal{F}^{-1}(\varphi_p(\cdot))$. Therefore: $u_p(\cdot) \in \mathcal{S}(\mathbb{R}^n) \ \forall p = 1, 2, \ldots$, and $\hat{u}_p(\xi) = \varphi_p(\xi), \ \forall \xi \in \mathbb{R}^n$.

We now have, using (5.3), the sequence of estimates

$$\int_{|\xi|\geq 1} |\varphi_p(\xi)|^2 d\xi \leq C \int_{\mathbb{R}^n} (1 + |\xi|^2)^r |\varphi_p(\xi)|^2 d\xi, \ \forall p \in \mathbb{N}, \tag{5.4}$$

and, accordingly:

$$\int_{1\leq|\xi|\leq p} (1 + |\xi|^2)^{-n/2} d\xi \leq C \int_{|\xi|\leq 2p} (1 + |\xi|^2)^r |\varphi_p(\xi)|^2 d\xi, \ \forall p \in \mathbb{N}. \tag{5.5}$$

The integral in the left-hand side $\to +\infty$ as $p \to \infty$; the integral in the right-hand side is estimated by

$C \int_{|\xi|\leq 2p} (1+|\xi|^2)^r (1+|\xi|^2)^{-n/2} d\xi$ which is convergent as $p \to \infty$ (due to $r < 0$).

We obtained a contradiction; this proves the Proposition 5.1.

Our subsequent result is stated as

Proposition 5.2 *Let again $\psi(\cdot) \in C^\infty(\mathbb{R}^n), \psi = 0$ for $|\xi| \leq \frac{1}{2}, \psi = 1$ for $|\xi| \geq 1, 0 \leq \psi(\cdot) \leq 1$; let $\sigma \in \mathbb{R}$, and $\psi_\sigma(\xi) = \psi(\xi)|\xi|^\sigma$ for $\xi \neq 0, \psi_\sigma(\xi) = 0$ for*

$\xi = 0$. *Consider the operator* $\psi_\sigma(D) = \mathcal{F}^{-1}\mathcal{M}_{\psi_\sigma}.\mathcal{F}, H^\infty \to H^\infty$. *Then we have :* $t.o\psi_\sigma(D) = \sigma$.

Proof We first note the estimate : $|\psi_\sigma(\xi)| \leq C(1+|\xi|^2)^{\sigma/2}$, $\forall \xi \in \mathbb{R}^n$ (in fact, we have, if $\sigma > 0$, $\left(\frac{|\xi|^2}{1+|\xi|^2}\right)^{\frac{\sigma}{2}} \leq 1$, hence $|\psi_p(\xi)| \leq (1+|\xi|^2)^{\sigma/2}$; $\forall \xi \in \mathbb{R}^n$; if $\sigma < 0$, note first that $\psi_p(\xi) = 0$ for $|\xi| \leq \frac{1}{2}$; then, for $|\xi| \geq \frac{1}{2}$ we have

$$\left(\frac{|\xi|^2}{1+|\xi|^2}\right)^{\frac{\sigma}{2}} = \left(\frac{1+|\xi|^2}{|\xi|^2}\right)^{\frac{|\sigma|}{2}} = \left(\frac{1}{|\xi|^2}+1\right)^{\frac{|\sigma|}{2}} \leq 5^{\frac{|\sigma|}{2}}$$

thus, we obtain $|\psi_p(\xi)| \leq 5^{\frac{|\sigma|}{2}}(1+|\xi|^2)^{\sigma/2}$, $\forall \xi \in \mathbb{R}^n$).

We now refer again to (1.14) and find that $\sigma \in \mathcal{O}((\psi_\sigma(D))$, hence $t.o(\psi_\sigma(D)) \leq \sigma$.

If the inequality is strict and $t.o(\psi_\sigma(D)) = \sigma_1 < \sigma$ we consider the operators $\psi_\sigma(D)$ and $\psi_{-\sigma}(D), H^\infty \to H^\infty$.

Now apply Proposition 4.2 (where $V = H^\infty, L_1 = \psi_\sigma(D), L_2 = \psi_{-\sigma}(D)$). We obtain

$$t.o(\psi_\sigma(D).\psi_{-\sigma}(D)) \leq t.o(\psi_\sigma(D)) + t.o(\psi_{-\sigma}(D)) = \sigma_1 + t.o(\psi_{-\sigma}) \leq \sigma_1 - \sigma < 0$$

Let us compute the product operator: $\psi_\sigma(D).\psi_{-\sigma}(D), H^\infty \to H^\infty$.

We have:

$\psi_\sigma(D)\psi_{-\sigma}(D) = \mathcal{F}^{-1}\mathcal{M}_{\psi_\sigma(\cdot)}\mathcal{F}.\mathcal{F}^{-1}\mathcal{M}_{\psi_{-\sigma(\cdot)}}\mathcal{F} = \mathcal{F}^{-1}\mathcal{M}_{\psi_\sigma}.\mathcal{M}_{\psi_{-\sigma}}.\mathcal{F} = \mathcal{F}^{-1}\mathcal{M}_{\psi^2}.\mathcal{F} = \psi^2(D)$.

Next, let us note that the function $\psi^2(\xi)$ has exactly the same properties (as specified in Prop. 5.1) as $\psi(\xi)$. Therefore $\psi^2(D)$ has true order $= 0$.

We obtained the false proposition $0 < 0$. This establishes Prop. 5.2.

Chapter II
Asymptotic expansions of linear operators in some vector spaces

Introduction

We consider the same general situation as in previous Chapter: a scale of Banach spaces $(E^s)_{-\infty<s<\infty}$ over the same scalar field, satisfying (1.1)–(1.2), their intersection E^∞, and a vector subspace of it, V. We then define and study some simple properties of *"asymptotic expansions"* and of *"p-equivalence"*, establishing mutual relationships.

First definitions and properties

Let $(r_j)_0^\infty$ be a strictly decreasing sequence of real numbers, such that:

$$\lim_{k\to\infty} r_k = -\infty. \tag{1.1}$$

Then, let $(A_j)_0^\infty$ be a sequence of operators in $\mathcal{L}in(V)$, such that:

$$t.o(A_j) = r_j, j = 0,1,2,\dots. \tag{1.2}$$

Definition 1.1 *The operator $M \in \mathcal{L}in(V)$ has asymptotic expansion $(A_j)_0^\infty$: $M \sim \sum_{j=0}^\infty A_j$, if the following sequence of strict inequalities holds :*

$$t.o(M - \sum_{j=0}^{N} A_j) < r_N, \ \forall N = 0,1,2,\dots \tag{1.3}$$

We note the following:

Proposition 1.1 *If $M \sim \sum_{j=0}^\infty A_j$, then*

$$t.o(M) = t.o(A_0) = r_0. \tag{1.4}$$

Proof Let us write the obvious equality: $M = M - A_0 + A_0$. Use Prop. 4.6 (Ch.I) and obtain:

$$t.o(M) = t.o((M - A_0) + A_0) \leq \max(t.o(M - A_0),\ t.o(A_0)). \tag{1.5}$$

From (1.3) it follows: $t.o(M - A_0) < r_0$. We also have, $t.o(A_0) = r_0$.

Hence, $\max(t.o(M - A_0), t.o(A_0)) = r_0$, and accordingly

$$t.o(M) \leq r_0. \tag{1.6}$$

Next, let us write the obvious equality: $A_0 = (A_0 - M) + M$.

Again, we find that

$$t.o(A_0) \leq \max(t.o(A_0 - M),\ t.o(M)). \tag{1.7}$$

On the other hand (Prop. 4.3, Ch.I) we know that: $t.o(A_0 - M) = t.o(M - A_0) < r_0$;

If we assume that the inequality in (1.6) is strict (that is $t.o(M) < r_0$), we derive from (1.7) that $t.o(A_0) < r_0$, a contradiction. Therefore we obtain: $t.o(M) = r_0$. ∎

Next, we consider the subalgebra: $\mathcal{E} = \mathcal{L}in_{\mathcal{O}\neq\phi}(V)$ (see Section 2 - Ch.I). We give

Definition 1.2 *The operators A, B in $\mathcal{E}$ are "p-equivalent"* (notation: $A \overset{p}{\sim} B$) *iff*

$$t.o(A - B) \leq \min(t.o(A), t.o(B)) \tag{1.8}$$

with strict inequality when $t.o(A) \in \mathbb{R}$, $t.o(B) \in \mathbb{R}$.

Proposition 1.2 *The "p-equivalence" is a reflexive relation: $A \overset{p}{\sim} A \forall A \in \mathcal{E}$.*

Proof We have $A - A = \Theta$, hence $t.o(A - A) = -\infty$. If $t.o(A) = r \in \mathbb{R}$, the corresponding inequality (1.8) is strict.

Proposition 1.3 *The "p-equivalence" is a symmetric property:*

$$A \overset{p}{\sim} B \Leftrightarrow B \overset{p}{\sim} A. \tag{1.9}$$

Proof We know (Prop. 4.3, Ch.I), that : $t.o(A - B) = t.o(B - A)$. Hence (1.8) follows, for both operators $A - B$ and $B - A$.

We shall now establish

Proposition 1.4 *If $A, B \in \mathcal{E}$ are p-equivalent, then*

$$t.o(A) = t.o(B). \tag{1.10}$$

Proof We have the equality : $A = (A - B) + B$. Therefore, again

$$t.o(A) \leq \max(t.o(A - B), t.o(B)). \tag{1.11}$$

On the other hand we see that (from (1.8))

$$t.o(A - B) \leq t.o(B),$$

Hence : $t.o(A) \leq t.o(B)$

Use also Prop. 1.3, and find : $t.o(B) \leq t.o(A)$. Hence, their equality follows. ∎

Let us see now that the p-equivalence has also the "transivity" property, and is, accordingly, a true "equivalence relation".

Proposition 1.5 *Let* $A, B, C \in \mathcal{E}$ *and* $A \overset{p}{\sim} B, B \overset{p}{\sim} C$. *Then* $A \overset{p}{\sim} C$.

Proof We use the (obvious) equality : $A - C = (A - B) + (B - C)$

Therefore (again by Prop. 4.6 – Ch.I) we have

$$t.o(A - C) \leq \max(t.o(A - B),\ t.o(B - C)) \tag{1.12}$$

On the other hand, from previous Proposition (1.4) we get : $t.o(A) = t.o(B) = t.o(C)$.

We now consider two cases:

i) $t.o(A - B) \leq t.o(B - C)$. It follows that (using (1.12)

$t.o(A - C) \leq t.o(B - C) \leq \min(t.o(B), t.o(C))$ with strict inequality when the true order of B and C are finite (this last inequality from (1.8)).

As $t.o(B) = t.o(A)$ we have also

$t.o(A - C) \leq \min(t.o(A), t.o(C))$, with strict inequality in the finite case.

This means precisely that $A \overset{p}{\sim} C$.

ii) The second case will arise when: $t.o(B - C) \leq t.o(A - B)$.

It follows (again from (1.12)), that

$t.o(A - C) \leq t.o(A - B) \leq \min(t.o(A), t.o(B)) = \min(t.o(A), t.o(C))$

with strict inequality in the finite case. Again $A \overset{p}{\sim} C$.

∎

Of importance is the

Proposition 1.6 *Let $A, B \in \mathcal{E}$ and $t.o(A - B) \leq t.o(B)$ (strict inequality when $t.o(A - B) > -\infty$). Then $A \overset{p}{\sim} B$.*

Proof Again we write the equality: $A = (A - B) + B$

We have then : $t.o(A) \leq \max(t.o(A - B), t.o(B)) = t.o(B)$.

Let us assume first that : $t.o(A - B) = -\infty$.

Write $B = (B - A) + A$; then $t.o(B) \leq \max(t.o(A - B), t.o(A)) = t.o(A)$.

Hence, in this case ($t.o(A - B) = -\infty$) we get : $t.o(A) \leq t.o(B) \leq t.o(A)$, hence $t.o(A) = t.o(B)$.

Next, let us assume : $t.o(A - B) > -\infty$ (hence $t.o(A - B) < t.o(B)$). Again we have : $t.o(A) \leq t.o(B)$. Assume that $t.o(A) < t.o(B)$ (strictly). As previously : $t.o(B) \leq \max(t.o(A - B), t.o(A))$. Now we have $t.o(A - B) < t.o(B)$ and $t.o(A) < t.o(B)$ (both strict inequalities). Therefore, $\max(t.o(A - B), t.o(A)) < t.o(B)$ (strictly), and we arrived at the contradiction : $t.o(B) < t.o(B)$.

Therefore, $t.o(A) = t.o(B)$.

Thus, in any case, $t.o(A) = t.o(B)$, and so we get:

$t.o(A - B) \leq t.o(B) = t.o(A) = \min(t.o(A), t.o(B)), A \overset{p}{\sim} B$.

■

Proposition 1.7 *If $A, B \in \mathcal{E}$ and $A \overset{p}{\sim} B$ then $-A \overset{p}{\sim} -B$*

Proof We have : $-A - (-B) = B - A$ and $t.o(B - A) = t.o(A - B)$.

■

Proposition 1.8 *If $A \in \mathcal{E}$ and $A \overset{p}{\sim} \Theta$, then $t.o(A) = -\infty$.*

Proof We have $t.o(A - \Theta) = t.o(A) \leq \min(t.o(A), t.o(\Theta)) = -\infty$.

■

Proposition 1.9 *If $A, B \in \mathcal{E}$ and $t.o(A) = t.o(B) = -\infty$, then $A \overset{p}{\sim} B$.*

Proof We have $t.o(A - B) \leq \max(t.o(A), t.o(-B)) = \max(t.o(A), t.o(B)) = -\infty$. Then: $t.o(A - B) \leq \min(t.o(A), t.o(B)) = -\infty$, and $A \overset{p}{\sim} B$.

■

Finally, let us give

Proposition 1.10 *If $A \in \mathcal{E}$ and $t.o(B) = -\infty$, then $A + B \overset{p}{\sim} A$.*

Proof We have : $t.o(A + B - A) = t.o(B) = -\infty \leq \min(t.o(A + B), t.o(A))$.

■

2. Asymptotic expansions and p-equivalence

In this section we shall establish several connections between the previously discussed concepts of asymptotic expansion and of p-equivalence.

First we state

Proposition 2.1 *Let $M \in \mathcal{E}$ and then $(A_k)_0^\infty$ a sequence in $\mathcal{E}$, with the following properties:*

$$t.o(A_k) = r_k > -\infty \; ; \; M - (\sum_{j=0}^{k-1} A_j) \overset{p}{\sim} A_k \; ; \; \forall k = 1,2,\ldots ; \; M \overset{p}{\sim} A_0. \qquad (2.1)$$

Then the real sequence $(r_k)_0^\infty$ is strictly decreasing.

Proof From the relation : $M - A_0 \overset{p}{\sim} A_1$ and Prop. 1.4 we derive

$$t.o(M - A_0) = t.o(A_1) = r_1. \qquad (2.2)$$

Then, from relation : $M \overset{p}{\sim} A_0$ we derive (Definition 1.2), that $t.o(M - A_0) \leq \min(t.o(M), t.o(A_0))$, with possible strict inequality.

Here, $t.o(M) = t.o(A_0) = r_0 > -\infty$. We have accordingly the *strict* inequality:

$$t.o(M - A_0) < r_0. \qquad (2.3)$$

From (2.2) – (2.3) we then derive

$$r_1 < r_0 \qquad (2.4)$$

Next, we shall establish the strict inequality :

$$r_k > r_{k+1} \; , \; \forall k = 0,1,2,\ldots \qquad (2.5)$$

From the p-equivalence in (2.1) : $M - (A_0 + A_1 + \ldots A_{k-1}) \overset{p}{\sim} A_k$ and Prop. 1.4, we obtain : $t.o(M - \sum_0^{k-1} A_j) = r_k$; also we have : $t.o(M - \sum_0^k A_j) = t.o(A_{k+1}) = r_{k+1}$. Furthermore, using again the p-equivalence in (2.1) and definition (1.8) we obtain that $t.o(M - \sum_0^{k-1} A_j - A_k) \leq \min(t.o(M - \sum_0^{k-1} A_j), t.o(A_k)) = r_k$, and the inequality is strict. Thus, $r_{k+1} = t.o(M - \sum_0^{k-1} A_j - A_k) < r_k$.

■

Corollary *Assume that* : $\lim_{k\to\infty} r_k = -\infty$. *Then we have the asymptotic expansion* $M \sim \sum_0^\infty A_j$.

Proof From above Proposition we get : $t.o(M - \sum_0^{k-1} A_j) = t.o(A_k) = r_k < r_{k-1}$, $\forall k = 1, 2, \ldots$; this, together with : $r_k \to -\infty$, means precisely $M \sim \sum_0^\infty A_j$.

■

Let us note also the (simple)

Proposition 2.2 *Let* $M \in \mathcal{E}$ *and* $M \sim \sum_0^\infty A_k$. *Then* $M \overset{p}{\sim} A_0$ *and moreover,* $M - (\sum_0^{k-1} A_j) \overset{p}{\sim} A_k$, $\forall k = 1, 2, \ldots$.

Proof First we derive, from (1.3), that : $t.o(M - A_0) < r_0 = t.o(A_0)$. Hence, from, Prop. 1.6, we obtain that $M \overset{p}{\sim} A_0$ (and also $t.o(M) = r_0$). Next, again from (1.3), we get : $t.o(M - A_0 - A_1) < r_1 = t.o(A_1)$; from Prop. 1.6 we derive : $M - A_0 \overset{p}{\sim} A_1$.

And so on; for instance, from the inequality $t.o(M - A_0 - A_1 - \cdots - A_k) < r_k = t.o(A_k)$ we find that

$$M - A_0 - A_1 - \cdots - A_{k-1} \overset{p}{\sim} A_k.$$

■

Corollary From Prop. 2.2 and Prop. 1.4 we derive
$t.o(M - \sum_0^{k-1} A_j) = r_k = t.o(A_k)$ (hence, $t.o(M - A_0) = t.o(A_1)$, $t.o(M - A_0 - A_1) = t.o(A_2)$, etc.
This completes Prop. 1.1.

3. Asymptotic expansions-sequel

In this section we first establish a uniqueness result for asymptotic expansions, "modulo" operators of true order $= -\infty$. Precisely, the following is true.

Proposition 3.1 *Let* $\mathcal{F} \subset \mathcal{E}$ *enjoying the following property*

$$A, B \in \mathcal{F} \text{ and } A \overset{p}{\sim} B \Rightarrow t.o(A - B) = -\infty \tag{3.1}$$

Assume that $M \in \mathcal{E}$ *has two asymptotic expansions:*

$$M \sim \sum_0^\infty A_j \text{ and } M \sim \sum_0^\infty B_j \text{ where } A_j, B_j \in \mathcal{F}, j = 0, 1, 2, \ldots$$

Then : $t.o(A_j) = t.o(B_j)\ \forall j = 0, 1, 2 \ldots$ *and* $A_j - B_j \in \mathcal{L}_{-\infty} = \mathcal{L}in_{\mathcal{O}=\mathbb{R}}(V)$.

Proof First we use Prop. 2.2 and obtain : $M \overset{p}{\sim} A_0, M \overset{p}{\sim} B_0$. It follows that $A_0 \overset{p}{\sim} B_0$, and accordingly $t.o(A_0) = t.o(B_0)$. Also, from assumption, we have also that $t.o(A_0 - B_0) = -\infty$.

Next, again from Prop. 2.2, we derive

$M - A_0 \overset{p}{\sim} A_1, M - B_0 \overset{p}{\sim} B_1$. We also note that : $M - A_0 \overset{p}{\sim} M - B_0$ (in fact, their difference is $B_0 - A_0$ of true order $-\infty$).

Therefore $A_1 \overset{p}{\sim} M - A_0 \overset{p}{\sim} M - B_0 \overset{p}{\sim} B_1$ and $A_1 \overset{p}{\sim} B_1, t.o(A_1) = t.o(B_1), A_1 - B_1 \in \mathcal{L}_{-\infty}$.

Let us assume now that : $t.o(A_j) = t.o(B_j)\ \forall j = 0, 1, 2, \ldots, n$ and also that $A_j - B_j \in \mathcal{L}_{-\infty}\ \forall j = 0, 1, 2, \ldots, n$.

Use again Prop. 2.2 and obtain

$$M - (A_0 + A_1 + \cdots A_n) \overset{p}{\sim} A_{n+1}, M - (B_0 + B_1 + \cdots B_n) \overset{p}{\sim} B_{n+1}, \tag{3.2}$$

Also, note the obvious equality :

$M - (A_0 + A_1 + \cdots A_n) - (M - (B_0 + B_1 + \cdots B_n)) = (B_0 - A_0) + (B_1 - A_1) + \ldots + (B_n - A_n)$. From Prop. 2.2 –Ch.I, we deduce that

$$\sum_0^n (B_j - A_j) \in \mathcal{L}_{-\infty}.$$

Accordingly:

$$M - \sum_0^n A_j \overset{p}{\sim} M - \sum_0^n B_j \tag{3.3}$$

Hence, using (3.2) we obtain

$$A_{n+1} \overset{p}{\sim} M - \sum_0^n A_j \overset{p}{\sim} M - \sum_0^n B_j \overset{p}{\sim} B_{n+1} \text{ and } A_{n+1} \overset{p}{\sim} B_{n+1}.$$

$$\text{Therefore, } t.o(A_{n+1}) = t.o(B_{n+1}) \text{ and } A_{n+1} - B_{n+1} \in \mathcal{L}_{-\infty}. \tag{3.4}$$

The result is established now for any $n \in \mathbb{N}$.

∎

In next result we show that two operators in $\mathcal{E}$ with the same asymptotic expansion are equal modulo $\mathcal{L}_{-\infty}$.

Proposition 3.2 *Let* $M, N \in \mathcal{E}$ *and :* $M \sim \sum_0^\infty A_k, N \sim \sum_0^\infty A_k$. *Then* $t.o(M-N) = -\infty$.

Proof Using the definition of the asymptotic expansion ((1.3) – in Section 1), we find

$$t.o(M - \sum_0^n A_j) < r_n,\ t.o(N - \sum_0^n A_j) < r_n.$$

Next we write :
$M - N = (M - \sum_0^n A_j) - (N - \sum_0^n A_j)$. From Prop. 4.6 (Ch.I) we derive

$$t.o(M-N) \le \max\,(t.o(M - \sum_0^n A_j),\ t.o(-N + \sum_0^n A_j)) < r_n.$$

This is true $\forall n \in \mathbb{N}$, and we know also that $r_n \to -\infty$ as $n \to \infty$. Therefore, the result follows.

■

Proposition 3.3 *Let* $M \in \mathcal{E}$, $M \sim \sum_0^\infty A_k$ *and* $\lambda \in K/\{0\}$.
Then we have

$$\lambda M \sim \sum_0^\infty \lambda A_k \tag{3.6}$$

Proof We consider the difference : $\lambda M - \sum_0^n(\lambda A_k) = \lambda(M - \sum_0^n A_k)$.
Using Prop. 4.3 (Ch.I) we find

$$t.o(\lambda M - \sum_0^n(\lambda A_k)) = t.o(M - \sum_0^n A_k) < r_n.$$

Also, $t.o(\lambda A_k) = t.o(A_k) = r_k$. Hence, the result.

■

4. Finite expansions

It is possible to define an expansion in the finite case.

Let $r_0 > r_1 > \ldots > r_n$ a finite, strictly decreasing set of real numbers.

Let $A_0, A_1, \ldots, A_n$ be operators in $\mathcal{L}in(V)$, such that: $t.o(A_j) = r_j$, $j = 0, 1, \ldots, n$.

We say that the operator $M \in \mathcal{L}in(V)$ has the finite expansion $M \overset{f}{\sim} A_0 + A_1 + \ldots A_n$ iff

$$t.o(M - \sum_0^k A_j) < r_k, \ \forall k = 0, 1, \ldots, n. \tag{4.1}$$

We note the simple extension of Prop. 4.6 (Ch.I) :

Proposition 4.1 *If $L_1, L_2, \ldots L_n$ are operators in $\mathcal{L}in(V)$, then*

$$t.o(L_1 + L_2 + \ldots L_n) \leq \max(t.o(L_1), t.o(L_2), \ldots t.o(L_n)). \tag{4.2}$$

Proof We shall use induction on n. Thus, assume (4.2). Then (by Prop. 4.6 – Ch.I)

$t.o(L_1 + L_2 + \ldots L_{n+1}) \leq \max(t.o(L_1 + \ldots L_n), t.o(L_{n+1})) \leq$
$\max(\max(t.o(L_1), t.o(L_2), \ldots t.o(L_n)), t.o(L_{n+1})) = \max(t.o(L_1), t.o(L_2), \ldots t.o(L_{n+1}))$

■

Now, we shall be able to establish

Proposition 4.2 *Let $A_0, A_1, \ldots A_n \in \mathcal{L}in(V), t.o(A_j) = r_j$, $r_0 > r_1 > \ldots > r_n$, and $M = A_0 + A_1 + \ldots A_n$.*

Then $M \overset{f}{\sim} A_0 + A_1 + \ldots A_n$.

Proof We first have: $M - A_0 = A_1 + \ldots A_n$, hence $t.o(M - A_0) \leq \max (t.o(A_1), t.o(A_2), \ldots t.o(A_n)) = r_1 < r_0$.

Also, $M - (A_0 + A_1) = A_2 + A_3 + \ldots A_n$, hence
$t.o(M - (A_0 + A_1)) \leq \max(t.o(A_2), t.o(A_3), \ldots t.o(A_n)) = r_2 < r_1$.

In the same manner we find all inequalities (4.1).

Remark *The converse result is not true.* For instance, $M \overset{f}{\sim} A_0 \not\Rightarrow M = A_0$.

(This is seen as follows : $M \overset{f}{\sim} A_0$ means $t.o(M - A_0) < t.o(A_0)$).

Take $A_0 = \psi(D)$, the operator in Prop. 5.1 – Ch.I, acting in $V = H^\infty$.

We have : $r_0 = 0$.

Take $M = \psi_1(D)$ where $\psi_1(\xi) = 1 \ \forall \xi \in \mathbb{R}^n$. Hence $M =$ Identity in V.

It follows that

$M - A_0 = \eta(D)$ where $\eta(\xi) = \psi_1(\xi) - \psi(\xi) = 1$ for $|\xi| \leq \frac{1}{2}$, and $= 0$ for $|\xi| \geq 1$, $0 \leq \eta(\xi) \leq 1$.

Therefore, as $\eta(\xi)$ has compact support in $\mathbb{R}^n, t.o(\eta(D)) = -\infty$.

However $M \neq A_0$. [$\psi(D)$ is not the identity operator in H^∞ : otherwise we would have :

$(\psi(D)u)\hat{}(\xi) = \psi(\xi)\hat{u}(\xi) = \hat{u}(\xi)\ \forall u \in \mathcal{S},\ (\psi(\xi) - 1)\hat{u}(\xi) = 0\ \forall \xi \in \mathbb{R}^n,\ \forall u \in \mathcal{S}$.
(This is false : we have $\psi(\xi) - 1 = -1$ for $|\xi| \leq \frac{1}{2}$; take $\hat{u} \in C_0^\infty(\mathbb{R}^n), \hat{u}(\xi) = 1$ for $|\xi| \leq \frac{1}{2}$)].

■

Chapter III
Pseudo-differential operators in the spaces $\mathcal{B}_{1,s}(\mathbb{R}^n)$

Introduction

Pseudo-differential operators associated to a numerical valued symbol $p(x,\xi)$ were discussed in our monograph [9] (see especially Ch. 5 and Ch. 10). One formula for such an operator appears as follows:
$(\mathcal{P}(x,D)u)(x) = (2\pi)^{-n/2}\int_{\mathbb{R}^n} e^{i<x,\xi>}p(x,\xi)\hat{u}(\xi)d\xi$, where $x \in \mathbb{R}^n$ and $u \in \mathcal{S}(\mathbb{R}^n)$.

In the present chapter we discuss a class of such operators where the domain of definition is the space $\mathcal{B}_{1,s}(\mathbb{R}^n)$ (s a real number) (definition and simple properties of these spaces can be found in [9] - Ch. 12). Essentially (in somewhat different guise) the results in this section can be found in Friedrichs [4].

1. First definitions and properties

Let us start this section with the case of a linear partial differential operator of the form

$$\mathcal{G}(x,D) = \sum_{0\leq|\alpha|\leq m} g_\alpha(x)D^\alpha \tag{1.1}$$

where $x = (x_1, x_2 \ldots x_n) \in \mathbb{R}^n, D^\alpha = D_1^{\alpha_1}D_2^{\alpha_2}\ldots D_n^{\alpha_n}$; $\alpha = (\alpha_1,\alpha_2\ldots\alpha_n) \in \underline{\mathbb{N}}^n$; $D_j = \frac{1}{\sqrt{-1}}\frac{\partial}{\partial x_j}$; $g_\alpha(x) = g_{\alpha_1\alpha_2\ldots\alpha_n}(x_1,x_2,\ldots x_n)$ are complex-valued functions on $\mathbb{R}^n$; $|\alpha| = \alpha_1+\alpha_2+\ldots\alpha_n$.

Let $T \in \mathcal{B}_{1,m} = \{T \in \mathcal{S}'(\mathbb{R}^n), \hat{T} \in L^1_{loc}(\mathbb{R}^n) \text{ and } (1+|\xi|^2)^{m/2}\hat{T}(\xi) \in L^1(\mathbb{R}^n)\}$
(where $m \in \underline{\mathbb{N}}$). (1.2)

As seen in [9], if $T \in \mathcal{B}_{1,m}$ it follows that $T \in C^m(\mathbb{R}^n)$ and all the partial derivatives $\partial^\alpha T$, $0 \leq |\alpha| \leq m$, are bounded (continuous) functions on $\mathbb{R}^n$.

It follows that the expression

$$\mathcal{G}(x,D)T = \sum_{0\leq|\alpha|\leq m} g_\alpha(x)D^\alpha T \tag{1.3}$$

belongs to $C^o(\mathbb{R}^n)$ provided (say) that the coefficients $g_\alpha(x)$ are continuous over $\mathbb{R}^n$, and $T \in \mathcal{B}_{1,m}(\mathbb{R}^n)$

(even more : if $g_\alpha(x)$ are continuous and bounded over $\mathbb{R}^n$, then it results that $\mathcal{G}(x,D)T$ is also continuous and bounded over $\mathbb{R}^n$.

On the other hand, we can use the relation $(D^\alpha T)\hat{}(\xi) = \xi^\alpha \hat{T}(\xi)$, $|\alpha| \leq m$ to obtain (for $T \in \mathcal{B}_{1,m}$) that $\xi^\alpha \hat{T}(\xi)$, hence also $(D^\alpha T)\hat{}(\xi)$ belong to $L^1(\mathbb{R}^n)$ for $|\alpha| \leq m$. Using the Fourier inversion formula in $\mathcal{S}'(\mathbb{R}^n)$ sense (which here amounts to the Fourier transform in the L^1-classical sense) we obtain the relations

$$(D^\alpha T)(x) = (2\pi)^{-n/2} \int_{\mathbb{R}^n} e^{i<x,\xi>}(\xi^\alpha \hat{T}(\xi))d\xi, \ |\alpha| \leq m, \ x \in \mathbb{R}^n \qquad (1.4)$$

It follows that the representation formula

$$(\mathcal{G}(x,D)T)(x) = (2\pi)^{-n/2} \int_{\mathbb{R}^n} e^{i<x,\xi>}\left(\sum_{0\leq|\alpha|\leq m} g_\alpha(x)\xi^\alpha\right)\hat{T}(\xi)d\xi \qquad (1.5)$$

holds true.

As was done in [9] - Ch. 5, we may call the function $g(x,\xi) = \sum_{0\leq|\alpha|\leq m} g_\alpha(x)\xi^\alpha$, from $\mathbb{R}^n \times \mathbb{R}^n$ into $\mathbb{C}$, the (polynomial) symbol of the (pseudo)-differential operator $\mathcal{G}(x,D)$, and we have accordingly the usual representation formula of a pseudo-differential operator in terms of its symbol:

$$(\mathcal{G}(x,D)T)(x) = (2\pi)^{-n/2} \int_{\mathbb{R}^n} e^{i<x,\xi>} g(x,\xi)\hat{T}(\xi)d\xi, \ x \in \mathbb{R}^n \qquad (1.6)$$

where $T \in \mathcal{B}_{1,m}$

Next, we shall restrict ourselves to the space $\mathcal{B}_{1,0} = \mathcal{F}^{-1}(L^1(\mathbb{R}^n))$ and to the multiplication operator $\mathcal{M}_a$ where a is a function in $\mathcal{B}_{1,0}$ too. Thus we have:

$$T \in \mathcal{B}_{1,0} \to (\mathcal{M}_a T)(x) = a(x)T(x) \qquad (1.7)$$

We shall first see that : *The operator $T \to aT$ maps $\mathcal{B}_{1,0}$ into itself.*

Proof In fact, we have $\hat{a}(\xi) \in L^1(\mathbb{R}^n)$, and, if $U \in \mathcal{B}_{1,0}, \hat{U}(\xi) \in L^1(\mathbb{R}^n)$ too, so that

$$a(x) = (2\pi)^{-n/2} \int_{\mathbb{R}^n} e^{i<x,\xi>}\hat{a}(\xi)d\xi \text{ and } U(x) = (2\pi)^{-n/2} \int_{\mathbb{R}^n} e^{i<x,\xi>}\hat{U}(\xi)d\xi,$$

$$\forall x \in \mathbb{R}^n. \qquad (1.8)$$

As well-known, the ordinary convolution between $\hat{a}$ and $\hat{U}$, that is the function

$$(\hat{a} * \hat{U})(\xi) = \int_{\mathbb{R}^n} \hat{a}(\xi - \eta)\hat{U}(\eta)d\eta \text{ is defined a.e. on } \mathbb{R}^n$$

belongs to $L^1(\mathbb{R}^n)$ and moreover, the inequality

$$\| \hat{a} * \hat{U} \|_{L^1(\mathbb{R}^n)} \leq \| \hat{a} \|_{L^1(\mathbb{R}^n)} \cdot \| \hat{U} \|_{L^1(\mathbb{R}^n)} \text{ is verified.} \tag{1.9}$$

Furthermore, it is immediate that $(2\pi)^{-n/2} \int_{\mathbb{R}^n} e^{i<x,\xi>}(\hat{a} * \hat{U})(\xi)d\xi = (2\pi)^{n/2}a(x)U(x)$; thus: $aU \in \mathcal{F}^{-1}(L^1(\mathbb{R}^n))$ and $a.U \in \mathcal{B}_{1,0}$.

■

Remark It is now obvious that the operator $\mathcal{M}a : U \to aU$ from $\mathcal{B}_{1,0}$ into itself is a *linear* operator.

It is also a *continuous* operator, as follows from the estimate

$$\| a.U \|_{\mathcal{B}_{1,0}} = \int_{\mathbb{R}^n} |(a.U)^\wedge(\xi)|d\xi = (2\pi)^{-n/2} \int_{\mathbb{R}^n} |(\hat{a} * \hat{U})(\xi)|d\xi \leq (2\pi)^{-n/2}$$

$$\| \hat{a} \|_{L^1} . \| \hat{U} \|_{L^1} = \| a \|_{\mathcal{B}_{1,0}} (2\pi)^{-n/2} \| U \|_{\mathcal{B}_{1,0}}, \quad \forall U \in \mathcal{B}_{1,0} \tag{1.10}$$

Accordingly, for the operator norm of $\mathcal{M}_a$ we get the estimate

$$\| \mathcal{M}_a \|_{\mathcal{L}(\mathcal{B}_{1,0},\mathcal{B}_{1,0})} \leq \| a \|_{\mathcal{B}_{1,0}} . (2\pi)^{-n/2}. \tag{1.11}$$

Note the representation formula – obtained above – :

$$(\mathcal{M}_a U)(x) = (2\pi)^{-n} \int_{\mathbb{R}^n} e^{i<x,\xi>} \left(\int_{\mathbb{R}^n} \hat{a}(\xi - \eta)\hat{U}(\eta)d\eta \right) d\xi, \ \forall x \in \mathbb{R}^n \tag{1.12}$$

Now we introduce (following the pattern of partial differential operators – (see again [9] - Ch. 5) an operator called $\beta(D)$ associated to a symbol $\beta(\xi)$, *acting again in the space* $\mathcal{B}_{1,0}$. If, for instance, we assume $\beta(\xi)$ to be a *bounded measurable function* on $\mathbb{R}^n$, we see that if $\hat{U}(\xi) \in L^1(\mathbb{R}^n)$, then $\beta(\xi)\hat{U}(\xi) \in L^1(\mathbb{R}^n)$ too.

Then: *the formula*

$$(\beta(D)U)(x) = (2\pi)^{-n/2} \int_{\mathbb{R}^n} e^{i<x,\xi>} \beta(\xi)\hat{U}(\xi)d\xi, \ \forall x \in \mathbb{R}^n \tag{1.13}$$

defines a linear continuous operator on $\mathcal{B}_{1,0}(\mathbb{R}^n)$.

Proof In fact, $\beta(D)U$ is the inverse Fourier transform of the integrable function $\beta(\xi)\hat{U}(\xi)$, hence it belongs to $\mathcal{B}_{1,0}(\mathbb{R}^n)$. The linearity of the operator is again obvious, and then we have the simple estimates:

$$\| \beta(D)U \|_{\mathcal{B}_{1,0}} = \int_{\mathbb{R}^n} |(\beta(D)U)^\wedge(\xi)| d\xi = \int_{\mathbb{R}^n} |\beta(\xi)\hat{U}(\xi)| d\xi \leq \operatorname*{ess\,sup}_{\mathbb{R}^n} |\beta(\xi)| \, \| U \|_{\mathcal{B}_{1,0}}$$

so that we obtain the inequality

$$\| \beta(D) \|_{\mathcal{L}(\mathcal{B}_{1,0})} \leq \operatorname*{ess\,sup}_{\mathbb{R}^n} |\beta(\xi)| \tag{1.14}$$

giving an estimate of the operator norm of $\beta(D)$ through the essential supremum of its symbol.

Continuing in this vein, we shall now derive formulas for the product operators: $\mathcal{M}_a.\beta(D)$ and also $\beta(D)\mathcal{M}_a$.

We shall assume that $a \in \mathcal{B}_{1,0}$ and that $\beta = \beta(\xi)$ is a bounded measurable function on $\mathbb{R}^n$. Using previous results we get the representation formulas:

$$(\mathcal{M}_a\beta(D)U)(x) = (2\pi)^{-n} \int_{\mathbb{R}^n} e^{i<x,\xi>} \Big(\int_{\mathbb{R}^n} \hat{a}(\xi - \eta) (\beta(D)U)^\wedge(\eta) d\eta \Big) d\xi$$

$$= (2\pi)^{-n/2} \int_{\mathbb{R}^n} e^{i<x,\xi>} \Big((2\pi)^{-n/2} \int_{\mathbb{R}^n} \hat{a}(\xi - \eta)\beta(\eta)\hat{U}(\eta) d\eta \Big) d\xi, \ \forall U \in \mathcal{B}_{1,0} \quad \forall x \in \mathbb{R}^n \tag{1.15}$$

and also – for the reversed product:

$$(\beta(D)\mathcal{M}_a U)(x) = (2\pi)^{-n/2} \int_{\mathbb{R}^n} e^{i<x,\xi>} \beta(\xi)(\mathcal{M}_a U)^\wedge(\xi) d\xi =$$

$$(2\pi)^{-n/2} \int_{\mathbb{R}^n} e^{i<x,\xi>} \beta(\xi) \Big((2\pi)^{-n/2} \int_{\mathbb{R}^n} \hat{a}(\xi - \eta)\hat{U}(\eta) d\eta \Big) d\xi =$$

$$(2\pi)^{-n} \int_{\mathbb{R}^n} e^{i<x,\xi>} \Big(\int_{\mathbb{R}^n} \hat{a}(\xi - \eta)\beta(\xi)\hat{U}(\eta) d\eta \Big) d\xi, \ \forall U \in \mathcal{B}_{1,0} \quad \forall x \in \mathbb{R}^n. \tag{1.16}$$

Following the pattern which appears in (1.15)–(1.16), we are now defining in correspondence to a more general class of "symbols" $g(x,\xi)$, $\mathbb{R}^n \times \mathbb{R}^n \to \mathbb{C}$, two operators ("pseudo-differential"), acting as linear continuous mappings of $\mathcal{B}_{1,0}$ into itself.

We assume that $g(x,\xi)$ is defined as the inverse *partial* Fourier transform

$$g(x,\xi) = (2\pi)^{-n/2} \int_{\mathbb{R}^n} e^{i<x,\lambda>} \gamma(\lambda,\xi) d\lambda \tag{1.17}$$

where $\gamma(\lambda,\xi)$ is a measurable function on $\mathbb{R}^n \times \mathbb{R}^n$ (complex-valued), such that

$$\gamma(\lambda,\xi) \text{ is } \lambda\text{-measurable}, \ \forall \xi \in \mathbb{R}^n \tag{1.18}$$

$$\text{and } \ |\gamma(\lambda,\xi)| \leq k(\lambda) \ \forall \xi \in \mathbb{R}^n, \ \forall \lambda \in \mathbb{R}^n, \tag{1.19}$$

$$\text{where } k(\lambda) \in L^1(\mathbb{R}^n).$$

Remark From (1.18)–(1.19) it follows that: $\int_{\mathbb{R}^n} |\gamma(\lambda,\xi)| d\lambda \leq \int_{\mathbb{R}^n} k(\lambda) d\lambda = \| k \|_{L^1}$, $\forall \xi \in \mathbb{R}^n$.

We now define (at least formally) two (presumably different) operators corresponding to such a symbol $g(x,\xi)$.

The first one, denoted (as in (1.1)) with $\mathcal{G}(x,D)$ is given by the formula

$$(\mathcal{G}(x,D)U)(x) = \mathcal{F}^{-1}\left[(2\pi)^{-n/2} \int_{\mathbb{R}^n} \gamma(\xi-\eta,\eta)\hat{U}(\eta)d\eta\right], \ \forall x \in \mathbb{R}^n, \ \forall U \in \mathcal{B}_{1,0}(\mathbb{R}^n) \tag{1.20}$$

while the second one, now denoted with $G(x,D)$ is defined by the relation

$$(G(x,D)U)(x) = \mathcal{F}^{-1}\left[(2\pi)^{-n/2} \int_{\mathbb{R}^n} \gamma(\xi-\eta,\xi)\hat{U}(\eta)d\eta\right], \ \forall x \in \mathbb{R}^n, \ \forall U \in \mathcal{B}_{1,0}(\mathbb{R}^n) \tag{1.21}$$

In order to make sense of the integrals appearing in (1.20)–(1.21), we prove:

Lemma 1.1 *The function $B_1(\xi,\eta) = \gamma(\xi-\eta,\xi)$ is measurable with respect to η, for all $\xi \in \mathbb{R}^n$.*

Proof By (1.18), $\forall \xi \in \mathbb{R}^n$, the function $\mathbb{R}^n \to \mathbb{C}$: $\lambda \to \gamma(\lambda,\xi)$ is measurable. If (say), $\gamma(\lambda,\xi)$ is real valued, we get, $\forall \alpha \in \mathbb{R}$, that the set in $\mathbb{R}^n_\lambda$: $\mathcal{E}_{\alpha,\xi} = \{\lambda; \gamma(\lambda,\xi) > \alpha\}$ is measurable. Let $\mathcal{F}_{\alpha,\xi} = \{\eta, \gamma(\xi-\eta,\xi) > \alpha\}$.

We see that $\eta \in \mathcal{F}_{\alpha,\xi} \Leftrightarrow \xi - \eta \in \mathcal{E}_{\alpha,\xi} \Leftrightarrow -\eta \in -\xi + \mathcal{E}_{\alpha,\xi} \Leftrightarrow \eta \in \xi - \mathcal{E}_{\alpha,\xi}$. Hence $\mathcal{F}_{\alpha,\xi} = \xi - \mathcal{E}_{\alpha,\xi}$; as $\mathcal{E}_{\alpha,\xi}$ is measurable, so is $\mathcal{F}_{\alpha,\xi}$.

■

Lemma 1.2 *The function $B_2(\xi,\eta) = \gamma(\xi-\eta,\eta)$ is measurable with respect to η for almost all $\xi \in \mathbb{R}^n$.*

Proof We assumed that $\gamma(\lambda, t)$ is measurable in $\mathbb{R}^n \times \mathbb{R}^n$. We make the linear substitution: $\lambda = \xi - \eta, t = \eta$. The corresponding function $B_2(\xi, \eta)$ is also measurable in $\mathbb{R}^n \times \mathbb{R}^n$. Even more: $B_2(\xi, \eta)$ is integrable on $\{\xi, |\xi| \leq p\} \times \mathbb{R}^n (\forall p \in \mathbb{N})$: in fact, using (1.19) – we obtain: $|B_2(\xi, \eta)| = |\gamma(\xi - \eta, \eta)| \leq k(\xi - \eta)$, $\forall (\xi, \eta) \in \mathbb{R}^\times \mathbb{R}^n$; hence the iterated integral: $\int_{|\xi| \leq p} (\int_{\eta \in \mathbb{R}^n} |B_2(\xi, \eta)| d\eta) d\xi$ is estimated by $\int_{|\xi| \leq p} (\int_{\eta \in \mathbb{R}^n} k(\xi - \eta) d\eta) d\xi = \Gamma_p \parallel k(\cdot) \parallel_{L^1(\mathbb{R}^n)}$. Using Tonelli's theorem we find that the $2n$-integral: $\int_{|\xi| \leq p} \int_{\eta \in \mathbb{R}^n} |B_2(\xi, \eta)| d\xi d\eta$ exists and is $\leq \Gamma_p \parallel k(\cdot) \parallel_{L^1(\mathbb{R}^n)}$. Therefore, from Fubini's theorem we find that the function $\eta \to B_2(\xi, \eta)$ is integrable (hence measurable) on $\mathbb{R}^n$, for almost all ξ in the ball $\{\xi \in \mathbb{R}^n, |\xi| \leq p\}$. Taking balls of radius $p = 1, 2, 3, \ldots$, we find that $\eta \to B_2(\xi, \eta)$ is measurable on $\mathbb{R}^n$, for almost all $\xi \in \mathbb{R}^n$.

■

Consider now the integral in (1.20). The expression under the integral sign is accordingly $\gamma(\xi - \eta, \eta) \hat{U}(\eta) = B_2(\xi, \eta) \hat{U}(\eta)$, where $\hat{U}(\eta) \in L^1(\mathbb{R}^n)$.

Using Lemma 1.2 we find that, for almost all $\xi \in \mathbb{R}^n$, the function $\eta \to B_2(\xi, \eta) \hat{U}(\eta)$ is measurable. We also have the estimate: $|B_2(\xi, \eta) \hat{U}(\eta)| \leq k(\xi - \eta) |\hat{U}(\eta)|$ and $\int_{\mathbb{R}^n_\eta} k(\xi - \eta) |\hat{U}(\eta)| d\eta = H(\xi) \in L^1(\mathbb{R}^n)$. Therefore, the function $\eta \to B_2(\xi, \eta) \hat{U}(\eta)$ belongs to $L^1(\mathbb{R}^n)$ for almost all $\xi \in \mathbb{R}^n$ and $\int_{\mathbb{R}^n} \gamma(\xi - \eta, \eta) \hat{U}(\eta) d\eta$ as a function of ξ, belongs to $L^1(\mathbb{R}^n_\xi)$.

(We can see this fact in a slightly different way, as follows: consider the $2n$-integral $\int \int_{\mathbb{R}^n \times \mathbb{R}^n} \gamma(\xi - \eta, \eta) \hat{U}(\eta) d\xi d\eta$. We see that the iterated integral:

$$\int_{\mathbb{R}^n} \int_{\mathbb{R}^n} (|\gamma(\xi - \eta, \eta)| \, |\hat{U}(\eta)| d\xi) d\eta =$$

$$\int_{\mathbb{R}^n} \left(\int_{\mathbb{R}^n} (|\gamma(\xi - \eta, \eta)| \, d\xi \right) |\hat{U}(\eta)| d\eta \leq \int_{\mathbb{R}^n} \left(\int_{\mathbb{R}^n} k(\xi - \eta) d\xi \right) |\hat{U}(\eta)| d\eta$$

$$= \parallel k \parallel_{L^1} \parallel \hat{U} \parallel_{L^1} < \infty$$

is absolutely convergent.

Hence, by Tonelli's theorem, the $2n$-integral $\int \int_{\mathbb{R}^n \times \mathbb{R}^n} \gamma(\xi - \eta, \eta) \hat{U}(\eta) d\xi d\eta$ is (absolutely) convergent; then, by Fubini's theorem we obtain that the integral: $\int_{\mathbb{R}^n} \gamma(\xi - \eta, \eta) \hat{U}(\eta) d\eta$ is integrable on $\mathbb{R}^n_\xi$).

This last result now shows that the (inverse) Fourier transform which appears in (1.20) is well-defined, in classical $L^1(\mathbb{R}^n)$-sense. Accordingly, we find from (1.20) that $\mathcal{G}(x, D)U$ is a continuous bounded function on $\mathbb{R}^n_x$ as well as a member of $\mathcal{B}_{1,0}(\mathbb{R}^n)$.

In a similar way, using Lemma 1.1 and the estimate

$$|\gamma(\xi-\eta,\xi)\hat{U}(\eta)| \leq k(\xi-\eta)|\hat{U}(\eta)|, \ \forall \xi \in \mathbb{R}^n, \ \eta \in \mathbb{R}^n$$

we find that

$G(x,D)U$ is also a continuous bounded function on $\mathbb{R}^n$ and a member of $\mathcal{B}_{1,0}(\mathbb{R}^n)$.

It is also obvious that both G and $\mathcal{G}$ are linear operators, $\mathcal{B}_{1,0}(\mathbb{R}^n) \to \mathcal{B}_{1,0}(\mathbb{R}^n)$ (it follows from definitions (1.20)–(1.21).

Remark Consider the special case:

$g(x,\xi) = a(x)b(\xi)$ where $a(\cdot) \in \mathcal{B}_{1,0}$ and $b(\cdot)$ is bounded measurable on $\mathbb{R}^n(|b(\xi)| \leq M \ \forall \xi \in \mathbb{R}^n)$.

Therefore $a(x) = (2\pi)^{-n/2} \int_{\mathbb{R}^n} e^{i<x,\lambda>}\hat{a}(\lambda)d\lambda$, $\hat{a}(\cdot) \in L^1(\mathbb{R}^n)$ and, accordingly $g(x,\xi) = (2\pi)^{-n/2} \int_{\mathbb{R}^n} e^{i<x,\lambda>}\hat{a}(\lambda)b(\xi)d\lambda$; thus $\gamma(\lambda,\xi) = \hat{a}(\lambda)b(\xi)$, which is measurable on $\mathbb{R}^n \times \mathbb{R}^n$; $\forall \xi \in \mathbb{R}^n$ we have $|b(\xi)| \leq M$, hence $|\hat{a}(\lambda)b(\xi)| \leq M|\hat{a}(\lambda)| \in L^1(\mathbb{R}^n)$. Thus (1.18)–(1.19) are verified, with $k(\lambda) = M|\hat{a}(\lambda)|$.

Computing $\mathcal{G}(x,D)$ in this case we find:

$\mathcal{G}(x,D)U = \mathcal{F}^{-1}\left[(2\pi)^{-n/2} \int_{\mathbb{R}^n} \hat{a}(\xi-\eta)b(\eta)\hat{U}(\eta)d\eta\right]$ – same as in (1.15).

Therefore

$\mathcal{G}(x,D)U = \mathcal{M}_{a(\cdot)}b(D)U$ (by (1.15)).

Also, we immediately see that

$G(x,D)U = \mathcal{F}^{-1}\left[(2\pi)^{-n/2} \int_{\mathbb{R}^n} \hat{a}(\xi-\eta)b(\xi)\hat{U}(\eta)d\eta\right]$ – same as in (1.16).

Therefore

$G(x,D)U = b(D)\mathcal{M}_{a(\cdot)}U$ (by (1.16)).

We end this section with the simple estimate for the operator norm of both operators $\mathcal{G}$ and G, from $\mathcal{B}_{1,0}$ into itself.

Using (1.20) we derive:

$$\| \mathcal{G}(x,D)U \|_{\mathcal{B}_{1,0}} = \| (2\pi)^{-n/2} \int_{\mathbb{R}^n} \gamma(\xi-\eta,\xi)\hat{U}(\eta)d\eta \|_{L^1(\mathbb{R}^n)} \leq$$

$$(2\pi)^{-n/2} \| k \|_{L^1(\mathbb{R}^n)} \cdot \| \hat{U} \|_{L^1} = (2\pi)^{-n/2} \| k \|_{L^1(\mathbb{R}^n)} \| U \|_{\mathcal{B}_{1,0}(\mathbb{R}^n)} .$$

A similar estimate holds for $G(x,D)U$.

Therefore, the operator norms of both operators $\mathcal{G}$ and G, from $\mathcal{B}_{1,0}$ into itself, are estimated (from above) by

$$(2\pi)^{-n/2} \| k \|_{L^1(\mathbb{R}^n)}$$

2. Operators in $\mathcal{B}_{1,s}(\mathbf{R}^n)$(where $s \in \mathbf{R}$)

A somewhat different situation is obtained by consideration of symbols $g(x,\xi)$ in a more general class:

We assume: $g(x,\xi)$ is a measurable function, $\mathbb{R}^n \times \mathbb{R}^n \to \mathbb{C}$, such that, for some real number r, we have the representation formula:

$$g(x,\xi).(1+|\xi|^2)^{-r/2} = (2\pi)^{-n/2} \int_{\mathbb{R}^n} e^{i<x,\lambda>}\gamma(\lambda,\xi)d\lambda, \ \forall\xi \in \mathbb{R}^n, \ \forall x \in \mathbb{R}^n \quad (2.1)$$

where:

$\gamma(\lambda,\xi)$ is measurable in $\mathbb{R}^n \times \mathbb{R}^n$; $\lambda \to \gamma(\lambda,\xi)$ is measurable $\forall\xi \in \mathbb{R}^n$;

$$\left.\begin{array}{c}\text{for } \xi \in \mathbb{R}^n,\ |\gamma(\lambda,\xi)| \leq k(\lambda), \text{ a (measurable) function such that} \\ (1+|\lambda|)^{|s|}k(\lambda) \in L^1(\mathbb{R}^n) \ \text{(for some real number } s) \end{array}\right\} \quad (2.2)$$

(therefore, the previous situation arises when $s = r = 0$).

If we introduce the notation:

$$\gamma_r(\lambda,\xi) = (1+|\xi|^2)^{r/2}\gamma(\lambda,\xi) \quad (2.3)$$

we obtain: $|\gamma_r(\lambda,\xi)| \leq (1+|\xi|^2)^{r/2}k(\lambda)$ and then, using (2.1), the equality:

$$g(x,\xi) = (2\pi)^{-n/2} \int_{\mathbb{R}^n} e^{i<x,\lambda>}\gamma_r(\lambda,\xi)d\lambda, \ \forall x \in \mathbb{R}^n, \ \forall\xi \in \mathbb{R}^n. \quad (2.4)$$

We define again operators $\mathcal{G}(x,D)$ and $G(x,D)$ on $\mathcal{B}_{1,s+r}$ and we prove that they are linear continuous mappings of this space into $\mathcal{B}_{1,s}$.

The formal defintions are as in (1.20)-(1.21), that is:

$$\mathcal{G}(x,D)U = \mathcal{F}^{-1}\big[(2\pi)^{-n/2} \int_{\mathbb{R}^n} \gamma_r(\xi-\eta,\eta)\hat{U}(\eta)d\eta\big], \quad (2.5)$$

$$G(x,D)U = \mathcal{F}^{-1}\big[(2\pi)^{-n/2} \int_{\mathbb{R}^n} \gamma_r(\xi-\eta,\xi)\hat{U}(\eta)d\eta\big], \quad (2.6)$$

where $U \in \mathcal{B}_{1,s+r}$, that is $(1+|\eta|^2)^{s+r/2}|\hat{U}(\eta)| \in L^1(\mathbb{R}^n)$.

In order to establish convergence of the integrals and the necessary estimates, we need an inequality, similar to (10.10) in [9], which here appears as

Lemma 2.1 *For any real number s and $\forall\xi,\eta \in \mathbb{R}^n$, the inequality*

$$(1+|\xi|^2)^{s/2} \leq (1+|\xi-\eta|)^{|s|}(1+|\eta|^2)^{s/2} \quad (2.7)$$

holds true.

Proof Applying the mean-value theorem we get:

$(1+|\xi|^2)^{1/2} - (1+|\eta|^2)^{1/2} = \langle \xi - \eta, grad(1+|\xi|^2)^{1/2}|_{\xi=\zeta}\rangle$ where $\langle\ \rangle$ is the scalar product in $\mathbb{R}^n$, while ζ is a point on the line segment $[\xi, \eta]$.

We also note that: $\frac{\partial}{\partial \xi_j}(1+|\xi|^2)^{1/2} = \frac{1}{2}(1+|\xi|^2)^{-1/2}.2\xi_j = \xi_j(1+|\xi|^2)^{-1/2}$, $\forall j = 1, 2, \ldots n$.

Accordingly we obtain:

$$\| grad(1+|\xi|^2)^{1/2} \|_{\mathbb{R}^n} = \Big(\sum_{j=1}^{n} \xi_j^2 (1+|\xi|^2)^{-1}\Big)^{1/2} = \Big(\frac{|\xi|^2}{1+|\xi|^2}\Big)^{1/2} \leq 1$$
$$\forall \xi \in \mathbb{R}^n.$$

It results therefore:

$$(1+|\xi|^2)^{1/2} - (1+|\eta|^2)^{1/2} \leq |\xi - \eta|, \ \forall \xi, \eta \in \mathbb{R}^n \tag{2.8}$$

Hence, we also get:

$$\begin{aligned}(1+|\xi|^2)^{1/2} &\leq (1+|\eta|^2)^{1/2} + |\xi-\eta| \leq (1+|\eta|^2)^{1/2} + |\xi-\eta|(1+|\eta|^2)^{1/2} \\ &= (1+|\eta|^2)^{1/2}(1+|\xi-\eta|).\end{aligned} \tag{2.9}$$

Now, if $s > 0$, (2.9) implies

$$(1+|\xi|^2)^{s/2} \leq (1+|\eta|^2)^{s/2}(1+|\xi-\eta|)^s, \ \forall \xi, \eta \in \mathbb{R}^n. \tag{2.10}$$

Next, if $s < 0$ we obtain from (2.9) that

$$(1+|\xi|^2)^{s/2} \geq (1+|\eta|^2)^{s/2}(1+|\xi-\eta|)^s, \ \forall \xi, \eta \in \mathbb{R}^n. \tag{2.11}$$

Changing ξ with η, (2.11) becomes
$(1+|\eta|^2)^{s/2} \geq (1+|\xi|^2)^{s/2}(1+|\xi-\eta|)^s$ which in turn implies that
$(1+|\xi|^2)^{s/2} \leq (1+|\eta|^2)^{s/2}(1+|\xi-\eta|)^{-s}$ which is again (2.7).

■

We shall now estimate the expressions under integral sign in (2.5) and (2.6). It is readily seen that

$$
\begin{aligned}
&|\gamma_r(\xi-\eta,\eta)\hat{U}(\eta)| \leq (1+|\eta|^2)^{r/2} k(\xi-\eta)|\hat{U}(\eta)| \\
&= k(\xi-\eta)(1+|\eta|^2)^{s+r/2}(1+|\eta|^2)^{-s/2}.|\hat{U}(\eta)| \\
&= k(\xi-\eta)(1+|\eta|^2)^{s+r/2}(1+|\xi|^2)^{s/2}(1+|\eta|^2)^{-s/2}(1+|\xi|^2)^{-s/2}|\hat{U}(\eta)| \quad (2.12)
\end{aligned}
$$

and also (using (2.7))

$$
|\gamma_r(\xi-\eta,\xi)\hat{U}(\eta)| \leq (1+|\xi|^2)^{r/2}k(\xi-\eta)|\hat{U}(\eta)| \leq (1+|\eta|^2)^{r/2}(1+|\xi-\eta|)^{|r|}k(\xi-\eta).
$$

$$
|\hat{U}(\eta)| \leq (1+|\eta|^2)^{s+r/2}|\hat{U}(\eta)|(1+|\eta|^2)^{-s/2}(1+|\xi-\eta|)^{|r|}k(\xi-\eta). \quad (2.13)
$$

We first continue with estimate (2.12) : (using (2.7)) we obtain that

$$
|\gamma_r(\xi-\eta,\eta)\hat{U}(\eta)| \leq k(\xi-\eta)(1+|\xi-\eta|)^{|s|}(1+|\xi|^2)^{-s/2}(1+|\eta|^2)^{s+r/2}|\hat{U}(\eta)| \quad (2.14)
$$

We next consider the integral of the right-hand side in (2.14), that is the expression

$$
\left(\int_{\mathbb{R}^n} k(\xi-\eta)(1+|\xi-\eta|)^{|s|}(1+|\eta|^2)^{s+r/2}|\hat{U}(\eta)|d\eta\right)(1+|\xi|^2)^{-s/2} = (1+|\xi|^2)^{-s/2}g(\xi)
$$

where $g(\cdot)$ is the convolution between the integrable functions on $\mathbb{R}^n$: $(1+|\lambda|)^{|s|}k(\lambda)$ and $(1+|\eta|^2)^{s+r/2}|\hat{U}(\eta)|$ (see (2.2) and use: $U \in \mathcal{B}_{1,s+r}$). Therefore $g(\xi)$ is defined a.e. on $\mathbb{R}^n$ and $g(\xi) \in L^1(\mathbb{R}^n)$.

As previously seen, the function $\gamma_r(\xi-\eta,\eta)\hat{U}(\eta)$ is measurable in $\eta \in \mathbb{R}^n$ and is estimated in absolute value by the right-hand side of (2.14). Therefore

$$
\left|\int_{\mathbb{R}^n} \gamma_r(\xi-\eta,\eta)\hat{U}(\eta)d\eta\right| \leq (1+|\xi|^2)^{-s/2}g(\xi), \text{ a.e. in } \mathbb{R}^n. \quad (2.15)
$$

We next prove also that the function $\xi \to \int_{\mathbb{R}^n}\gamma_r(\xi-\eta,\eta)\hat{U}(\eta)d\eta$ is $\mathbb{R}^n$-measurable. Consider the iterated integral:

$\int_{\mathbb{R}^n}(1+|\xi|^2)^{s/2}(\int_{\mathbb{R}^n}|\gamma_r(\xi-\eta,\eta)|\,|\hat{U}(\eta)|d\eta)d\xi$ which is $\leq$
$\int_{\mathbb{R}^n}(1+|\xi|^2)^{s/2}(1+|\xi|^2)^{-s/2}g(\xi)d\xi = \| g(\cdot)\|_{L^1(\mathbb{R}^n)}$. Thus, the $2n$-integral
$\int\int_{\mathbb{R}^n\times\mathbb{R}^n}(1+|\xi|^2)^{s/2}\gamma_r(\xi-\eta,\eta)\hat{U}(\eta)d\xi d\eta$ is absolutely convergent ;

this implies that $\Gamma(\xi) = \int_{\mathbb{R}^n}(1+|\xi|^2)^{s/2}\gamma_r(\xi-\eta,\eta)\hat{U}(\eta)d\eta$ belongs to $L^1(\mathbb{R}^n)$ and consequently, the function $\xi \to \int_{\mathbb{R}^n}\gamma_r(\xi-\eta,\eta)\hat{U}(\eta)d\eta$ which equals $(1+|\xi|^2)^{-s/2}\Gamma(\xi)$ is measurable.

Furthermore, the estimate (2.15) above indicates that the function $\int_{\mathbb{R}^n}\gamma_r(\xi-\eta,\eta)\hat{U}(\eta)d\eta$ defines a temperate distribution on $\mathbb{R}^n$.

Accordingly, the inverse Fourier transform in (2.5) is taken in $\mathcal{S}'(\mathbb{R}^n)$ sense. Then, obviously, $(\mathcal{G}(x,D)U)^\wedge$ equals $(2\pi)^{-n/2}\int_{\mathbb{R}^n}\gamma_r(\xi-\eta,\eta)\hat{U}(\eta)d\eta$ and $(1+|\xi|^2)^{s/2}(\mathcal{G}(x,D)U)^\wedge(\xi)$ belongs to $L^1(\mathbb{R}^n)$, which means that the distribution $\mathcal{G}(x,D)U$, where $U \in \mathcal{B}_{1,s+r}$, belongs to $\mathcal{B}_{1,s}(\mathbb{R}^n)$. Note also (from (2.5) again), that the mapping $\mathcal{G}(x,D)$ is linear between these two spaces.

Next, let us note the norm estimate

$$\| \mathcal{G}(x,D)U \|_{\mathcal{B}_{1,s}} = \| (1+|\xi|^2)^{s/2}(\mathcal{G}(x,D)U)^\wedge(\xi) \|_{L^1(\mathbb{R}^n)} \le$$
$$(2\pi)^{-n/2} \| g(\cdot) \|_{L^1(\mathbb{R}^n)} \le (2\pi)^{-n/2} \| (1+|\lambda|)^{|s|}k(\lambda) \|_{L^1(\mathbb{R}^n)} \| U \|_{\mathcal{B}_{1,s+r}}$$

as results from the definition of $g(\cdot)$ as a convolution.

We find therefore the upper bound for the operator norm of $\mathcal{G}(x,D)$:

$$\| \mathcal{G}(x,D) \|_{\mathcal{L}(\mathcal{B}_{1,s+r},\mathcal{B}_{1,s})} \le (2\pi)^{-n/2}\int_{\mathbb{R}^n}(1+|\lambda|)^{|s|}k(\lambda)d\lambda. \qquad (2.16)$$

Consider next, shortly, the operator $G(x,D)$ which has been defined by (2.6). Using (2.13) we see that

$$|\gamma_r(\xi-\eta,\xi)\hat{U}(\eta)| \le (1+|\xi-\eta|)^{|r|}k(\xi-\eta)(1+|\eta|^2)^{-s/2}(1+|\eta|^2)^{s+r/2}|\hat{U}(\eta)|$$
$$\le (1+|\xi-\eta|)^{|r|}k(\xi-\eta)(1+|\xi-\eta|)^{|s|/2}(1+|\xi|^2)^{-s/2}(1+|\eta|^2)^{s+r/2}|\hat{U}(\eta)|$$
$$= (1+|\xi-\eta|)^{|r|+(|s|/2)}k(\xi-\eta)(1+|\xi|^2)^{-s/2}(1+|\eta|^2)^{s+r/2}|\hat{U}(\eta)|$$

Let us make now the

New Assumption

The function $(1+|\lambda|)^{|r|+(|s|/2)}k(\lambda)$ *belongs to* $L^1(\mathbb{R}^n)$.

One obtains then, with similar reasonings as for $\mathcal{G}(x,D)$ that the operator $G(x,D)$ in (2.6) is a linear continuous mapping of $\mathcal{B}_{1,s+r}$ into $\mathcal{B}_{1,s}$.

3. The operator $\mathcal{G}(x, D)$ in $\mathcal{B}_{1,1}(\mathbb{R}^n)$

We consider symbols g defined by

$$g(x,\xi) = (2\pi)^{-n/2} \int_{\mathbb{R}^n} e^{i<x,\lambda>} \gamma(\lambda,\xi) d\lambda \tag{3.1}$$

where $\gamma(\lambda,\xi)$ verifies conditions (2.2) with $s = 1$.

It obviously results that

$$\frac{\partial g}{\partial x_j} = \mathcal{F}^{-1}[i\lambda_j \gamma(\lambda,\xi)],\ j = 1,2,\dots n \tag{3.2}$$

where the (classical) partial derivatives are continuous and bounded in $\mathbb{R}^n$.

Then, by section 2, the operator $\mathcal{G}(x,D)$ given by $\mathcal{F}^{-1}\big[(2\pi)^{-n/2} \int_{\mathbb{R}^n} \gamma(\xi-\eta,\eta)\hat{U}(\eta)d\eta\big]$ is a linear continuous mapping of $\mathcal{B}_{1,1}(\mathbb{R}^n)$ into itself and also of $\mathcal{B}_{1,0}(\mathbb{R}^n)$ into itself.

Next, using (3.2) and results in section 1 we find that the operator (denoted here) with $\mathcal{G}_{ix_j}(x,D)$ which is associated to the symbol $\frac{1}{\sqrt{-1}}\frac{\partial g}{\partial x_j}$ by (1.20), is a linear continuous mapping, $\mathcal{B}_{1,0} \to \mathcal{B}_{1,0}$.

Let us now prove

Proposition 3.1 *The identity*

$$\frac{1}{\sqrt{-1}}\frac{\partial}{\partial x_j}(\mathcal{G}(x,D)U) = \mathcal{G}(x,D)\Big(\frac{1}{\sqrt{-1}}\frac{\partial U}{\partial x_j}\Big) + \mathcal{G}_{ix_j}(x,D)U,\ \forall U \in \mathcal{B}_{1,1}(\mathbb{R}^n) \tag{3.3}$$

holds true.

Proof Remember that $\mathcal{B}_{1,1} \subset \mathcal{B}_{1,0}$; also, $U \in \mathcal{B}_{1,1} \Rightarrow \frac{\partial U}{\partial x_j} \in \mathcal{B}_{1,0}$ (for we have $(\frac{\partial U}{\partial x_j})^\wedge(\xi) = \sqrt{-1}\xi_j \hat{U}(\xi) \in L^1$ – due to $|\xi_j \hat{U}(\xi)| \leq |\xi|\ |\hat{U}(\xi)| \leq (1+|\xi|^2)^{1/2}|\hat{U}(\xi)|$).

Thus, the identity (3.3) makes sense in $\mathcal{B}_{1,0}(\mathbb{R}^n)$.

Now, in order to establish (3.3), we use the representation formula

$$\mathcal{G}(x,D)U = \mathcal{F}^{-1}\big[(2\pi)^{-n/2} \int_{\mathbb{R}^n} \gamma(\xi-\eta,\eta)\hat{U}(\eta)d\eta\big],\ U \in \mathcal{B}_{1,1} \tag{3.4}$$

Therefore (in $\mathcal{S}'(\mathbb{R}^n)$-sense), we obtain the equalities:

$$\frac{1}{i}\frac{\partial}{\partial x_j}(\mathcal{G}(x,D)U) = \mathcal{F}^{-1}\Big(\int_{\mathbb{R}^n} \xi_j \gamma(\xi-\eta,\eta)\hat{U}(\eta)d\eta\Big) =$$

$$\mathcal{F}^{-1}\Big(\int_{\mathbb{R}^n} (\xi_j - \eta_j)\gamma(\xi-\eta,\eta)\hat{U}(\eta)d\eta + \int_{\mathbb{R}^n} \gamma(\xi-\eta,\eta)\eta_j\hat{U}(\eta)d\eta\Big). \tag{3.5}$$

(the integrals make sense: $U \in \mathcal{B}_{1,1} \Rightarrow \eta_j \hat{U}(\eta) \in L^1$).

Next, note that : $\eta_j \hat{U}(\eta) = \mathcal{F}(\frac{1}{i}\frac{\partial U}{\partial x_j})$. Accordingly we get

$$\mathcal{F}^{-1}\left(\int_{\mathbb{R}^n} \gamma(\xi-\eta,\eta)\eta_j \hat{U}(\eta)d\eta\right) = \mathcal{G}(x,D)\left(\frac{1}{i}\frac{\partial U}{\partial x_j}\right) \tag{3.6}$$

Note also (from (3.2)) the relation: $\lambda_j \gamma(\lambda,\xi) = \mathcal{F}(\frac{1}{i}\frac{\partial g}{\partial x_j})$. It follows:

$$\int_{\mathbb{R}^n} (\xi_j - \eta_j)\gamma(\xi-\eta,\eta)\hat{U}(\eta)d\eta = \mathcal{F}(\mathcal{G}_{ix_j}(x,D)U). \tag{3.7}$$

Therefore, turning back to (3.5) we now have:

$$\frac{1}{i}\frac{\partial}{\partial x_j}(\mathcal{G}(x,D)U) = \mathcal{G}_{ix_j}U + \mathcal{G}(x,D)\left(\frac{1}{i}\frac{\partial U}{\partial x_j}\right), \text{ that is (3.3)}$$

■

Chapter IV
Pseudo-differential operators in $\mathcal{F}^{-1}(L^p)$ and in $H^s(\mathbb{R}^n)$ spaces

Introduction

In this chapter we discuss previously defined couple of operators $G(x,D)$ and $\mathcal{G}(x,D)$ in some different function spaces:

$$\mathcal{F}^{-1}(L^p(\mathbb{R}^n)) = \{T \in \mathcal{S}'(\mathbb{R}^n), \hat{T}(\xi) \in L^p(\mathbb{R}^n), p \geq 1\} \tag{0.1}$$

$$H^s(\mathbb{R}^n) = \{T \in \mathcal{S}'(\mathbb{R}^n), (1+|\xi|^2)^{s/2}\hat{T}(\xi) \in L^2(\mathbb{R}^n)\} \tag{0.2}$$

Again, simple assumptions on the symbol $g(x,\xi)$ ensure the continuity of above operators in these spaces. One could consult reference [15] for a similar – but slightly different – presentation.

1. Operators in $\mathcal{F}^{-1}(L^p(\mathbb{R}^n))$

The definition of the space indicated with $\mathcal{F}^{-1}(L^p(\mathbb{R}^n))$ extends that of the space $\mathcal{B}_{1,0} = \mathcal{F}^{-1}(L^1(\mathbb{R}^n))$.

Let us denote: $\mathcal{F}^{-1}(L^p) = X$

We see that for $p = 2$, in view of Plancherel's theorem, we have exactly $L^2(\mathbb{R}^n) = X$.

It is immediate that X is a vector space over $\mathbb{C}$; a norm is defined on X by the relation

$$\| T \|_{\mathcal{F}^{-1}(L^p)} = \| \hat{T} \|_{L^p} = \left(\int_{\mathbb{R}^n} |\hat{T}(\xi)|^p d\xi \right)^{1/p}, \quad \forall T \in X \tag{1.1}$$

We next see that X *is a complete (hence a Banach) space over* $\mathbb{C}$. Take therefore a Cauchy sequence in X, say $(U_j)_1^\infty$. It follows that the sequence of Fourier transforms: $(\hat{U}_j)_1^\infty$ is a Cauchy sequence in the (complete) space $L^p(\mathbb{R}^n)$. Therefore, there exists $f(\cdot) \in L^p(\mathbb{R}^n)$ such that

$$\int_{\mathbb{R}^n} |\hat{U}_j(\xi) - f(\xi)|^p d\xi \to 0 \text{ as } j \to \infty$$

It T_f is the temperate distribution associated to f, we define: $U = \mathcal{F}^{-1}(T_f)$. Thus $U \in X, \hat{U} = f$, and we have $\int_{\mathbb{R}^n} |\hat{U}_j(\xi) -$

$\hat{U}(\xi)|^p d\xi \quad \to \quad 0$, which means precisely that $U_j \quad \to \quad U$ in X-sense.

■

Next, let us note the following:

$$\mathcal{S}(\mathbb{R}^n) \subset L^p(\mathbb{R}^n); \text{ hence } \mathcal{F}^{-1}(\mathcal{S}) = \mathcal{S} \subset X, \ \forall p \geq 1.$$

Actually, $\mathcal{S}$ *is dense in* X: if $U \in X$, then $\hat{U} \in L^p$ and there exists a sequence $(\phi_j)_1^\infty$ where $\{\phi_j\}_1^\infty \subset C_0^\infty(\mathbb{R}^n)$, such that : $\phi_j \to \hat{U}$ in L^p-sense.

Let us put now: $\psi_j = \mathcal{F}^{-1}(\phi_j), j = 1, 2, \ldots$ Therefore, $\psi_j \in \mathcal{S}(\mathbb{R}^n)$ and :

$$\| U - \psi_j \|_X = \left(\int_{\mathbb{R}^n} |\hat{U}(\xi) - \phi_j(\xi)|^p d\xi \right)^{1/p} \to 0 \text{as } j \to \infty \tag{1.2}$$

■

We next define symbols $g(x, \xi)$ and the associated couple of pseudo-differential operators: $G(x, D)$ and $\mathcal{G}(x, D)$.

The definition of the symbol is slightly more general with respect to Ch. III as we now allow $g(x, \xi)$ to remain undefined for $\xi = 0$.

Thus: we consider measurable functions $g(x, \xi), \mathbb{R}^n \times \mathbb{R}^n/\{0\} \to \mathbb{C}$, admitting a representation formula

$$g(x, \xi) = (2\pi)^{-n/2} \int_{\mathbb{R}^n} e^{i<x,\lambda>} \gamma(\lambda, \xi) d\lambda, \ (x, \xi) \in \mathbb{R}^n \times \mathbb{R}^n/\{0\} \tag{1.3}$$

where $\gamma(\lambda, \xi)$ is a measurable function, $\mathbb{R}^n \times \mathbb{R}^n/\{0\} \to \mathbb{C}$, such that $\lambda \to \gamma(\lambda, \xi)$ is also measurable, $\mathbb{R}^n \to \mathbb{C}$, $\forall \xi \in 0$, and $|\gamma(\lambda, \xi)| \leq k(\lambda)$, $\forall \xi \in \mathbb{R}^n/\ \{o\}, \lambda \in \mathbb{R}^n$, where $k(\lambda) \in L^1(\mathbb{R}^n)$.

Taking $U \in X$ (thus $\hat{U} \in L^p$), we have the estimates

$$|\gamma(\xi - \eta, \xi)\hat{U}(\eta)| \leq k(\xi - \eta)|\hat{U}(\eta)| \tag{1.4}$$

$$|\gamma(\xi - \eta, \eta)\hat{U}(\eta)| \leq k(\xi - \eta)|\hat{U}(\eta)| \tag{1.5}$$

As $k(\cdot) \in L^1$ and $\hat{U} \in L^p$, if we apply Young's theorem (see for instance [6]) we obtain that the integral: $\int_{\mathbb{R}^n} k(\xi - \eta)|\hat{U}(\eta)| d\eta$, exists *a.e* and belongs to $L^p(\mathbb{R}^n)$; its p-norm is also estimated by $\| k \|_{L^1} \| \hat{U} \|_{L^p}$. In view of (1.4)–(1.5), it follows that the integrals

$$\int_{\mathbb{R}^n} \gamma(\xi - \eta, \xi)\hat{U}(\eta) d\eta, \ \int_{\mathbb{R}^n} \gamma(\xi - \eta, \eta)\hat{U}(\eta) d\eta \tag{1.6}$$

both define functions in $L^p(\mathbb{R}^n)$ which have their L^p-norm estimated by $\| k \|_{L^1} \| \hat{U} \|_{L^p}$.

We remember finally definitions (1.20)–(1.21); it results that both operators $G(x, D)$ and $\mathcal{G}(x, D)$ are linear continuous mappings, $X \to X$ with operator norm $\leq (2\pi)^{-n/2} \| k(\cdot) \|_{L^1(\mathbb{R}^n)}$.

2. Operators in $H^s(\mathbb{R}^n)$

We next consider the same couple of operators acting in spaces $H^s(\mathbb{R}^n)$ – where s is any real number (we refer to [9] for definitions and basic properties of these spaces). We consider symbols $g(x, \xi), \mathbb{R}^n \times \mathbb{R}^n/\{0\} \to \mathbb{C}$, defined by (1.3) where

$$|\gamma(\lambda, \xi)| \leq k(\lambda) \ \forall \xi \in \mathbb{R}^n/\{0\}, \text{ and } \int_{\mathbb{R}^n} (1 + |\lambda|^2)^{|s|/2} k(\lambda) d\lambda < +\infty \tag{2.1}$$

The formal definition of operators $G(x, D), \mathcal{G}(x, D)$ is again given in (1.20)–(1.21) - Ch. III. Now, as $U \in H^s(\mathbb{R}^n)$ we find that $(1 + |\eta|^2)^{s/2} \hat{U}(\eta) \in L^2(\mathbb{R}^n)$. Again we have estimates (1.4)–(1.5), and write also the relation

$$k(\xi - \eta)|\hat{U}(\eta)| = (1 + |\eta|^2)^{-s/2} k(\xi - \eta)(1 + |\eta|^2)^{s/2} |\hat{U}(\eta)| \tag{2.2}$$

Then we use inequality (10.10) in [9] in the form

$$(1 + |\eta|^2)^{-s/2} \leq 2^{|s|/2}(1 + |\eta - \xi|^2)^{|s|/2}(1 + |\xi|^2)^{-s/2}, \ \forall s \in \mathbb{R}, \xi, \eta \in \mathbb{R}^n \tag{2.3}$$

Then, from (2.2) and (2.3) we obtain the inequality

$$k(\xi-\eta)|\hat{U}(\eta)| \leq 2^{|s|/2}(1+|\xi|^2)^{-s/2}(1+|\xi-\eta|^2)^{|s|/2} k(\xi-\eta)(1+|\eta|^2)^{s/2}|\hat{U}(\eta)|, \ \forall \xi, \eta \in \mathbb{R}^n \tag{2.4}$$

Again from Young's theorem, using (2.1) and because $U \in H^s(\mathbb{R}^n)$, we obtain that

$$\int_{\mathbb{R}^n} (1 + |\xi - \eta|^2)^{|s|/2} k(\xi - \eta)(1 + |\eta|^2)^{s/2} |\hat{U}(\eta)| d\eta \quad \text{exists for almost all } \xi \in \mathbb{R}^n,$$

and is a function in $L^2(\mathbb{R}^n)$.

Using (2.4) we then obtain that

$$\int_{\mathbb{R}^n} k(\xi - \eta)|\hat{U}(\eta)| d\eta \text{ exists } a.e. \text{ on } \mathbb{R}^n,$$

and the inequality

$$\int_{\mathbb{R}^n} k(\xi-\eta)|\hat{U}(\eta)|d\eta \le 2^{|s|/2}(1+|\xi|^2)^{-s/2}$$
$$\int_{\mathbb{R}^n} (1+|\xi-\eta|^2)^{|s|/2}k(\xi-\eta)(1+|\eta|^2)^{s/2}|\hat{U}(\eta)|d\eta$$

holds true, *a.e.* (2.5)

Therefore: $(1+|\xi|^2)^{s/2}\int_{\mathbb{R}^n} k(\xi-\eta)|\hat{U}(\eta)|d\eta$ belongs to $L^2(\mathbb{R}^n)$. This obviously implies that: $K(\xi)=\int_{\mathbb{R}^n} k(\xi-\eta)|\hat{U}(\eta)|d\eta \in L^1_{loc}(\mathbb{R}^n)$ and $K(\xi)$ defines also a temperate distribution on $\mathbb{R}^n$.

Our next step: use the estimates : $|\gamma(\xi-\eta,\xi)\hat{U}(\eta)| \le k(\xi-\eta)|\hat{U}(\eta)|$ and $|\gamma(\xi-\eta,\eta)\hat{U}(\eta)| \le k(\xi-\eta)|\hat{U}(\eta)|$, in order to obtain that the functions of ξ, defined by the integrals

$$\int_{\mathbb{R}^n} \gamma(\xi-\eta,\xi)\hat{U}(\eta)d\eta \text{ and } \int_{\mathbb{R}^n} \gamma(\xi-\eta,\eta)\hat{U}(\eta)d\eta \tag{2.6}$$

belong to $L^1_{loc}(\mathbb{R}^n_\xi)$, and furthermore that, $(1+|\xi|^2)^{s/2}\int_{\mathbb{R}^n}\gamma(\xi-\eta,\xi)\hat{U}(\eta)d\eta$ and $(1+|\xi|^2)^{s/2}\int_{\mathbb{R}^n}\gamma(\xi-\eta,\eta)\hat{U}(\eta)d\eta$ are elements of $L^2(\mathbb{R}^n)$.

In turn, this shows that the functions in (2.6) define temperate distributions and their inverse Fourier transforms

$$\mathcal{F}^{-1}[(2\pi)^{-n/2}\int_{\mathbb{R}^n} \gamma(\xi-\eta,\xi)\hat{U}(\eta)d\eta],\ \mathcal{F}^{-1}[(2\pi)^{-n/2}\int_{\mathbb{R}^n} \gamma(\xi-\eta,\eta)\hat{U}(\eta)d\eta] \tag{2.7}$$

are elements of the space $H^s(\mathbb{R}^n)$; precisely, we have

$$G(x,D)U = \mathcal{F}^{-1}[(2\pi)^{-n/2}\int_{\mathbb{R}^n} \gamma(\xi-\eta,\xi)\hat{U}(\eta)d\eta],\ \forall U\in H^s(\mathbb{R}^n) \tag{2.8}$$

$$\mathcal{G}(x,D)U = \mathcal{F}^{-1}[(2\pi)^{-n/2}\int_{\mathbb{R}^n} \gamma(\xi-\eta,\eta)\hat{U}(\eta)d\eta],\ \forall U\in H^s(\mathbb{R}^n) \tag{2.9}$$

Note also the important estimate

$$\| \int_{\mathbb{R}^n} (1+|\xi-\eta|^2)^{|s|/2}k(\xi-\eta)(1+|\eta|^2)^{s/2}|\hat{U}(\eta)|d\eta \|_{L^2(\mathbb{R}^n)}$$
$$\le \| (1+|\lambda|^2)^{|s|/2}k(\lambda) \|_{L^1(\mathbb{R}^n)} \| U \|_{H^s(\mathbb{R}^n)}$$

(following from Young's theorem) and consequently

$$\| G(x,D)U \|_{H^s(\mathbb{R}^n)} = \| (1+|\xi|^2)^{s/2}(2\pi)^{-n/2} \int_{\mathbb{R}^n} \gamma(\xi-\eta,\xi)\hat{U}(\eta)d\eta \|_{L^2}$$
$$\leq 2^{|s|/2}(2\pi)^{-n/2} \| (1+|\lambda|^2)^{|s|/2}k(\lambda) \|_{L^1(\mathbb{R}^n)} \cdot \| U \|_{H^s(\mathbb{R}^n)} \tag{2.10}$$

and also

$$\| \mathcal{G}(x,D)U \|_{H^s(\mathbb{R}^n)} = \| (1+|\xi|^2)^{s/2}(2\pi)^{-n/2} \int_{\mathbb{R}^n} \gamma(\xi-\eta,\eta)\hat{U}(\eta)d\eta \|_{L^2}$$
$$\leq 2^{|s|/2}(2\pi)^{-n/2} \| (1+|\lambda|^2)^{|s|/2}k(\lambda) \|_{L^1(\mathbb{R}^n)} \cdot \| U \|_{H^s(\mathbb{R}^n)} \tag{2.11}$$

(see (2.4) – above).

Example *(Multiplication operator in* $H^s(\mathbb{R}^n)$ We consider the (very) special case of a symbol $g(x,\xi) = g(x)$ (independent of $\xi \in \mathbb{R}^n/\{0\}$), where $g(x) \in \mathcal{S}(\mathbb{R}^n)$. Accordingly we have

$$g(x) = (2\pi)^{-n/2} \int_{\mathbb{R}^n} e^{i<x,\lambda>}\hat{g}(\lambda)d\lambda \tag{2.12}$$

where $\hat{g}(\lambda) : (\mathcal{F}g)(\lambda))$ belongs to $\mathcal{S}(\mathbb{R}^n)$ as well.

The associated operators G and $\mathcal{G}$ are given by the formula

$$G(x,D)U = \mathcal{G}(x,D)U = \mathcal{F}^{-1}[(2\pi)^{-n/2} \int_{\mathbb{R}^n} \hat{g}(\xi-\eta)\hat{U}(\eta)d\eta)],\ \forall U \in H^s(\mathbb{R}^n).$$

If now $U \in \mathcal{S}(\mathbb{R}^n)$, then, by the classical result about Fourier transform of convolutions: $\hat{g} * \hat{U}$, where $\hat{g}, \hat{U} \in \mathcal{S}$, we obtain:
$G(x,D)U = \mathcal{G}(x,D)U = g(x).U(x) = \mathcal{M}g(\cdot).U$ where
$\mathcal{M}g(\cdot)$ is the multiplication operator by $g(\cdot)$.

Actually, as we now see, $\mathcal{M}g(\cdot)$ is, when applied to $U \in H^s(\mathbb{R}^n)$, the product (in $\mathcal{S}'(\mathbb{R}^n)$-sense), of g and U.

In fact, $\forall U \in \mathcal{S}'$, we know, when $g \in \mathcal{S}$, that $g \cdot U \in \mathcal{S}'$ is defined by

$$(gU)(\varphi) = U(g\varphi),\ \forall \varphi \in \mathcal{S}(\mathbb{R}^n) \tag{2.13}$$

On the other hand, we also know that $\mathcal{S}(\mathbb{R}^n)$ is dense in $H^s(\mathbb{R}^n)$ (in H^s-norm). Therefore, given $T \in H^s(\mathbb{R}^n), \exists (T_n)_1^\infty$, where $T_n \in \mathcal{S}(\mathbb{R}^n) \forall n \in \mathbb{N}$, and $T_n \to T$ in H^s-sense (hence, a fortiori, in $\mathcal{S}'$-sense).

We obtain, by the continuity property of operators G and $\mathcal{G}$ in H^s, that $\mathcal{G}(x,D)T = \lim_{n\to\infty} \mathcal{G}(x,D)T_n$ in H^s; hence, $\forall\varphi \in \mathcal{S}(\mathbb{R}^n)$,

$$(\mathcal{G}(x,D)T)(\varphi) = \lim_{n\to\infty} (\mathcal{G}(x,D)T_n)(\varphi) = \lim_{n\to\infty} (\mathcal{M}g(\cdot)T_n)(\varphi) =$$
$$\lim_{n\to\infty} (g(x)T_n(x))(\varphi) = \lim_{n\to\infty} T_n(g\varphi) = T(g\varphi) = (gT)(\varphi); \mathcal{G}(x,D)T = gT.$$

Thus, $\mathcal{G}(x,D)T$ (for $T \in H^s, g(x,\xi) = g(x) \in \mathcal{S}(\mathbb{R}^n)$) reduces to the multiplication (in $\mathcal{S}'$) between g and T. We can express this property also as follows:

If $g \in \mathcal{S}(\mathbb{R}^n), T \in H^s(\mathbb{R}^n)$, then $gT \in H^s(\mathbb{R}^n)$ and the estimate $\| gT \|_{H^s} \leq C_g \| T \|_{H^s}$, holds.

Extension (slightly more general symbols) The symbols $g(x,\xi)$ previously defined (by formulas (1.3)–(2.1)) have the immediate property that: $\lim_{|x|\to\infty} g(x,\xi) = 0 \; \forall\xi \in \mathbb{R}^n/\{0\}$ (it follows from $k(\cdot) \in L^1(\mathbb{R}^n)$).

We can dispense this restriction and consider instead symbols $g_1(x,\xi)$ of the form:

$$g_1(x,\xi) = g(x,\xi) + \tilde{g}(\xi) \tag{2.14}$$

where $g(x,\xi)$ is in the previous class while $\tilde{g}(\xi), \mathbb{R}^n/\{0\} \to \mathbb{C}$, is a bounded measurable function. Then, as seen previously (Ch. I – (1.10)), the operator $\tilde{g}(D) = \mathcal{F}^{-1}\mathcal{M}_{\tilde{g}(\cdot)}\mathcal{F}$ is a (well defined) linear continuous operator, $H^s(\mathbb{R}^n) \to H^s(\mathbb{R}^n)$, $\forall s \in \mathbb{R}$, and also $\mathcal{F}^{-1}(L^p) \to \mathcal{F}^{-1}(L^p), (p \geq 1)$, as readily seen. Then we define the two operators associated to $g_1(x,\xi)$ by the obvious relations

$$G_1(x,D) = G(x,D) + \tilde{g}(D), \mathcal{G}_1(x,D) = \mathcal{G}(x,D) + \tilde{g}(D), \tag{2.15}$$

and we now have again two linear continuous operators, $H^s(\mathbb{R}^n) \to H^s(\mathbb{R}^n)$, or $\mathcal{F}^{-1}(L^p) \to \mathcal{F}^{-1}(L^p), p \geq 1$.

(This extended situation permits us to include symbols $g_1(x,\xi) = \tilde{g}(\xi)$ which are independent of x; this is not possible when $g(x,\xi)$ must $\to 0$ as $|x| \to \infty$).

Chapter V
Alternative representation formulas for operators $G(x,D)$ and $\mathcal{G}(x,D)$

Introduction

In this short chapter we indicate a few different representation formulas for previously introduced couple of operators: $G(x,D)$ and $\mathcal{G}(x,D)$, which may appear useful for examination of various particular problems.

Operators on $\mathcal{S}(\mathbb{R}^n)$

We shall restrict the domain of definition of above indicated operators to the space $\mathcal{S}(\mathbb{R}^n)$ of rapidly decreasing $C^\infty(\mathbb{R}^n)$ functions, which is contained in all spaces: $\mathcal{B}_{1,s}(\mathbb{R}^n), \mathcal{F}^{-1}(L^p(\mathbb{R}^n))$ and $H^s(\mathbb{R}^n)$. We present

Proposition 1.1 *Let the symbol $g(x,\xi)$ be defined by (1.3) (Ch. IV). Then the representation formula*

$$G(x,D)u = \mathcal{F}^{-1}\left[(2\pi)^{-n/2}\int_{\mathbb{R}^n} e^{-i<x,\xi>} g(x,\xi)u(x)dx\right],\ \forall u \in \mathcal{S}(\mathbb{R}^n) \tag{1.1}$$

holds true.

Proof By assumptions, the symbol $g(x,\xi)$ is a continuous and bounded function on $\mathbb{R}^n_x$, for all $\xi \in \mathbb{R}^n/\{0\}$ fixed (we have, for instance, $|g(x,\xi)| \leq (2\pi)^{-n/2}\int_{\mathbb{R}^n} k(\lambda)d\lambda$, $\forall (x,\xi) \in \mathbb{R}^n \times \mathbb{R}^n/\{0\}$). Therefore, the integral in (1.1) is absolutely convergent for $u(\cdot) \in \mathcal{S}(\mathbb{R}^n)$.

Let us use now formula : $g(x,\xi) = (2\pi)^{-n/2}\int_{\mathbb{R}^n} e^{i<x,\lambda>}\gamma(\lambda,\xi)d\lambda$.

We obtain accordingly, $\forall u \in \mathcal{S}(\mathbb{R}^n)$, the relations:

$$\int_{\mathbb{R}^n} e^{-i<x,\xi>} g(x,\xi)u(x)dx = \int_{\mathbb{R}^n} e^{-i<x,\xi>}\left((2\pi)^{-n/2}\int_{\mathbb{R}^n} e^{i<x,\lambda>}\gamma(\lambda,\xi)d\lambda\right)u(x)dx$$

which equals (by Fubini's theorem) the absolutely convergent $(2n)$-integral

$$(2\pi)^{-n/2}\int_{\mathbb{R}^n\times\mathbb{R}^n} e^{-i<x,\xi-\lambda>}\gamma(\lambda,\xi)u(x)d\lambda dx = \int_{\mathbb{R}^n}\left(2\pi)^{-n/2}\int_{\mathbb{R}^n} e^{-i<x,\xi-\lambda>}u(x)dx\right)$$

$$\gamma(\lambda,\xi)d\lambda = \int_{\mathbb{R}^n}\hat{u}(\xi-\lambda).\gamma(\lambda,\xi)d\lambda = \int_{\mathbb{R}^n}\gamma(\xi-\eta,\xi)\hat{u}(\eta)d\eta.$$

Therefore, the expression in the right-hand side of (1.1) becomes now $\mathcal{F}^{-1}[(2\pi)^{-n/2}\int_{\mathbb{R}^n}\gamma(\xi-\eta,\xi)\hat{u}(\eta)d\eta]$, which is $G(x,D)u$.

We next have a similar result concerning operator $\mathcal{G}(x,D)$, expressed as

Proposition 1.2 *Under the same assumption on the symbol $g(x,\xi)$ we have the formula*

$$(\mathcal{G}(x,D)u(x) = (2\pi)^{-n/2}\int_{\mathbb{R}^n} e^{i<x,\eta>}g(x,\eta)\hat{u}(\eta)d\eta, \forall u\in\mathcal{S}(\mathbb{R}^n), \forall x\in\mathbb{R}^n \quad (1.2)$$

Proof First note the absolute convergence of the above integral, due to boundedness of g and absolute integrability of $\hat{u}$.

Next, the integral which appears in the definition of $\mathcal{G}(x,D)$: $\int_{\mathbb{R}^n}\gamma(\xi-\eta,\eta)\hat{u}(\eta)d\eta$, is a measurable function of ξ and is estimated, in absolute value, by the expression

$\int_{\mathbb{R}^n}k(\xi-\eta)|\hat{u}(\eta)|d\eta$, the convolution product of the integrable functions $k(\cdot)$ and $\hat{u}(\cdot)$. Thus, the function: $\xi\to\int_{\mathbb{R}^n}\gamma(\xi-\eta,\eta)\hat{u}(\eta)d\eta, (\forall u\in\mathcal{S})$, in an (absolutely) integrable function on $\mathbb{R}^n$, and, therefore, the expression $\mathcal{F}^{-1}[(2\pi)^{-n/2}\int_{\mathbb{R}^n}\gamma(\xi-\eta,\eta)\hat{u}(\eta)d\eta]$ (usually considered in the general, $\mathcal{S}'(\mathbb{R}^n)$-sense) here becomes the absolutely convergent Fourier integral

$(2\pi)^{-n/2}\int_{\mathbb{R}^n}e^{i<x,\xi>}((2\pi)^{-n/2}\int_{\mathbb{R}^n}\gamma(\xi-\eta,\eta)\hat{u}(\eta)d\eta)d\xi$ which equals (by application of Fubini's theorem)

$$(2\pi)^{-n/2}\int_{\mathbb{R}^n}[(2\pi)^{-n/2}\int_{\mathbb{R}^n}e^{i<x,\xi>}\gamma(\xi-\eta,\eta)d\xi]\hat{u}(\eta)d\eta = I$$

We now make, in the internal integral, the substitution: $\xi-\eta=\tau$, and obtain that

$$I = (2\pi)^{-n/2}\int_{\mathbb{R}^n}[(2\pi)^{-n/2}\int_{\mathbb{R}^n}e^{i<x,\eta+\tau>}\gamma(\tau,\eta)d\tau]\hat{u}(\eta)d\eta =$$

$$= (2\pi)^{-n/2}\int_{\mathbb{R}^n}[(2\pi)^{-n/2}\int_{\mathbb{R}^n}e^{i<x,\tau>}\gamma(\tau,\eta)d\tau]e^{i<x,\eta>}\hat{u}(\eta)d\eta =$$

$$(2\pi)^{-n/2}\int_{\mathbb{R}^n}g(x,\eta)e^{i<x,\eta>}\hat{u}(\eta)d\eta$$

This proves Prop. 1.2.

Our last result in this section deals with an alternative representation formula for $\mathcal{G}(x,D)$.

Proposition 1.3 *Let again, $g(x,\xi)$ be a measurable function, $\mathbb{R}^n \times \mathbb{R}^n/\{o\} \to \mathbb{C}$, given by: $g(x,\xi) = (2\pi)^{-n/2}\int_{\mathbb{R}^n} e^{i<x,\lambda>}\gamma(\lambda,\xi)d\lambda$, where $\gamma(\lambda,\xi)$ is also measurable, $\mathbb{R}^n \times \mathbb{R}^n/\{o\} \to \mathbb{C}, \gamma(\lambda,\xi)$ is λ-measurable $\forall\xi \in \mathbb{R}^n/\{o\}$, and $|\gamma(\lambda,\xi)| \leq k(\lambda) \in L^1(\mathbb{R}^n), \forall\lambda \in \mathbb{R}^n$, $\forall\xi \in \mathbb{R}^n/\{o\}$.*

Let $g_1(x,\xi) = (1+|\xi|^2)^{-m}g(x,\xi), \mathbb{R}^n \times \mathbb{R}^n/\{o\} \to \mathbb{C}$ and $g_2(x,y) = (2\pi)^{-n/2}\int_{\mathbb{R}^n} e^{i<\xi,y>}g_1(x,\xi)d\xi, \mathbb{R}^n \times \mathbb{R}^n \to \mathbb{C}$, where $m > n$.
Then the following holds:

$$(\mathcal{G}(x,D)u)(x) = (2\pi)^{-n/2}\int_{\mathbb{R}^n} g_2(x,x-y)[(I-\Delta)^m u](y)dy, \ \forall x \in \mathbb{R}^n, \ \forall u \in \mathcal{S}(\mathbb{R}^n) \tag{1.3}$$

Proof First we establish that the function $\xi \to g(x,\xi)$ is measurable on $\mathbb{R}^n/\{o\}, \forall x \in \mathbb{R}^n$. We note again that the measurable function $g(x,\xi)$ is also bounded:

$|g(x,\xi)| \leq (2\pi)^{-n/2}\int_{\mathbb{R}^n} k(\lambda)d\lambda, \ \forall(x,\xi) \in \mathbb{R}^n \times \mathbb{R}^n/\{o\}$

Hence, the $2n$-integral: $\int\int_{\substack{|x|\leq p\\ |\xi|\leq p}} g(x,\xi)dxd\xi$ is convergent.

It follows (th. of Fubini), that the function $\xi \in \mathbb{R}^n \to g(x,\xi)$ is measurable, for almost all $x \in \mathbb{R}^n, |x| \leq p$.

On the other hand, the function $x \to g(x,\xi)$ is continuous, as follows from the representation formula for g and the dominated convergence theorem.

Next, we can say that $\xi \to g(x,\xi)$ is measurable for $x \in \mathbb{R}^n, |x| \leq p, x \notin \mathcal{E}_p, m\mathcal{E}_p = 0$. Hence $\xi \to g(x,\xi)$ is measurable for $x \in \mathbb{R}^n, x \notin \bigcup_1^\infty \mathcal{E}_p$. If $\mathcal{E} = \bigcup_1^\infty \mathcal{E}_p$, then $m\mathcal{E} = 0$ and $\mathbb{R}^n/\mathcal{E}$ is dense in $\mathbb{R}^n$ – (otherwise, $\exists$ ball $S(\bar{x},\rho)$ contained in $\mathcal{E}$ and $m\mathcal{E} > 0$). Hence, if $x_0 \in \mathbb{R}^n, \exists(x_n)_1^\infty, x_n \in \mathbb{R}^n/\mathcal{E}\forall n = 1,2,\ldots, x_n \to x_0$. Then $g(x_0,\xi) = \lim_{n\to\infty} g(x_n,\xi)\forall\xi \in \mathbb{R}^n/\{o\}$. Thus $g(x_0,\xi)$ is ξ-measurable for all $x \in \mathbb{R}^n$.

It results that the function $g_1(x,\xi)$ is also ξ-measurable, $\forall x \in \mathbb{R}^n$. Furthermore: $|g_1(x,\xi)| \leq \sup_{\mathbb{R}^n\times\mathbb{R}^n/\{o\}} |g(x,\xi)|.(1+|\xi|^2)^{-m}$; hence the function $\xi \in \mathbb{R}^n/\{o\} \to g_1(x,\xi)$ belongs to $L^1(\mathbb{R}^n_\xi)$ (due to the condition $m > n$).

This permits the definition of $g_2(x,y)$ as an absolutely convergent Fourier integral, $\forall(x,y) \in \mathbb{R}^n \times \mathbb{R}^n$.

Now, we note that g_2 is a continuous function: if $(x_p,y_p) \to (x_0,y_0)$, then $e^{i<\xi,y_p>}g_1(x_p,\xi) \to e^{i<\xi,y_0>}g_1(x_0,\xi), \forall\xi \in \mathbb{R}^n/\{o\}$ and furthermore $|e^{i<\xi,y_p>}g_1(x_p,\xi)| \leq C.(1+|\xi|^2)^{-m} \in L^1(\mathbb{R}^n), \forall p = 1,2,\ldots$ (and apply the dominated convergence theorem).

Then, $\forall x \in \mathbb{R}^n$, the function $y \to g_2(x, x-y)$ is continuous too.

Also $g_2(x, x-y)$ is bounded: $|g_2(x, x-y)| \leq (2\pi)^{-n/2} \int_{\mathbb{R}^n} C.(1+|\xi|^2)^{-m} d\xi = C_1$. Thus, the integral in (1.3) is absolutely convergent.

To prove that it, in fact, defines $(\mathcal{G}(x, D)u)(x)$, we proceed as follows:

$$(2\pi)^{-n/2} \int_{\mathbb{R}^n} g_2(x, x-y)[(I-\Delta)^m u](y) dy =$$

$$(2\pi)^{-n/2} \int_{\mathbb{R}^n} ((2\pi)^{-n/2} \int_{\mathbb{R}^n} e^{i<\xi, x-y>} g_1(x,\xi) d\xi)[(I-\Delta)^m u](y) dy =$$

(after using Fubini's theorem again)

$$= (2\pi)^{-n/2} \int_{\mathbb{R}^n} ((2\pi)^{-n/2} \int_{\mathbb{R}^n} e^{-i<\xi, y>} [(I-\Delta)^m u](y) dy) e^{i<\xi, x>} g_1(x,\xi) d\xi =$$

$$= (2\pi)^{-n/2} \int_{\mathbb{R}^n} (1+|\xi|^2)^m \hat{u}(\xi).e^{i<\xi, x>} g_1(x,\xi) d\xi =$$

$$(2\pi)^{-n/2} \int_{\mathbb{R}^n} \hat{u}(\xi).e^{i<\xi, x>} g(x,\xi) d\xi$$

and apply Prop. 1.2.

■

Chapter VI
Kohn–Nirenberg homogeneous and C^∞-symbols and their associated operators

Introduction

Presently we shall explain the class of symbols proposed in the pioneering work of Kohn and Nirenberg [5], following essentially our expository paper [10]; we associate to these symbols a couple of operators A and $\mathcal{A}$ and establish some of their simple properties.

1. Symbols

The symbols in this Chapter are complex-valued functions $a(x,\xi)$ defined on $\mathbb{R}^n \times \mathbb{R}^n/\{o\}$, which are assumed to be C^∞-functions. Next, the positive homogeneity condition with respect to the ξ-variable

$$a(x,t\xi) = a(x,\xi)\forall(x,\xi) \in \mathbb{R}^n \times \mathbb{R}^n/\{o\},\ \forall t > 0 \tag{1.1}$$

holds true.

We also assume

$$\lim_{|x|\to\infty} a(x,\xi) = a(\infty,\xi) \text{ exists, } \forall\xi \in \mathbb{R}^n/\{o\} \tag{1.2}$$

$$\text{and } a(\infty,\xi) \in C^\infty(\mathbb{R}^n/\{o\}) \tag{1.3}$$

Denote: $a'(x,\xi) = a(x,\xi) - a(\infty,\xi), (x,\xi) \in \mathbb{R}^n \times \mathbb{R}^n/\{o\}$.

Thus:

$$a'(x,\xi) \in C^\infty(\mathbb{R}^n \times \mathbb{R}^n/\{o\}), \text{ and } a'(x,t\xi) = a'(x,\xi)\forall t > 0 \tag{1.4}$$

$x \in \mathbb{R}^n, \xi \in \mathbb{R}^n/\{o\}$

(in fact $a(\infty,t\xi) = \lim_{|x|\to\infty} a(x,t\xi) = \lim_{|x|\to\infty} a(x,\xi) = a(\infty,\xi), \forall t > 0, \xi \in \mathbb{R}^n/\{o\}$).

Finally we assume the sequence of estimates:

$$(1+|x|^2)^p|\partial_x^\alpha \partial_\xi^\beta a'(x,\xi)| \le C_{p,\alpha,\beta} \tag{1.5}$$

for $p \in \underline{\mathbb{N}}, \alpha \in \underline{\mathbb{N}}^n, \beta \in \underline{\mathbb{N}}^n, x \in \mathbb{R}^n, \xi \in \mathbb{R}^n/\{o\}$ and $|\xi| = 1$ (here $\underline{\mathbb{N}} = (0,1,2,\ldots) = \mathbb{N} \cup \{o\}$).

(The symbols ∂_x, ∂_ξ mean partial derivatives with respect to x or ξ respectively).

We can call these symbols, K-N symbols.

There are various consequences of the above described properties. First we prove

Proposition 1.1 *The function $a(\infty,\xi), \mathbb{R}^n/\{o\} \to \mathbb{C}$, satisfies the following inequality:*

$$|a(\infty,\xi) - a(\infty,\eta)| \leq C|\xi - \eta|(|\xi| + |\eta|)^{-1}, \ \forall \xi, \eta \in \mathbb{R}^n/\{o\}. \tag{1.6}$$

Proof Let $\xi, \eta \in \mathbb{R}^n/\{o\}$; put $\zeta = \frac{\xi}{|\xi|}, \mu = \frac{\eta}{|\eta|}$. It follows that $\zeta, \mu \in \mathbb{R}^n, |\zeta| = |\mu| = 1$; also $a(\infty,\zeta) = a(\infty,\xi)$ and $a(\infty,\mu) = a(\infty,\eta)$.

On the other hand we see that, for $\xi, \eta \in \mathbb{R}^n/\{o\}$, one has:

$$\frac{|\xi - \eta|}{|\xi| + |\eta|} = \frac{|\zeta|\xi| - \mu|\eta||}{|\xi| + |\eta|} = \left| \frac{|\xi|}{|\xi| + |\eta|}\zeta + \frac{|\eta|}{|\xi| + |\eta|}(-\mu) \right| \tag{1.7}$$

We now give the following (elementary)

Lemma 1.1 *If $0 \leq \theta \leq 1$, we have:*

$$|\theta\zeta + (1-\theta)(-\mu)| \geq \frac{1}{2}|\zeta - \mu| \tag{1.8}$$

for $|\zeta| = |\mu| = 1$.

Proof of Lemma We shall consider the expression

$g(\theta) = |\theta\zeta + (1-\theta)(-\mu)|^2 = \langle \theta\zeta + (1-\theta)(-\mu), \theta\zeta + (1-\theta)(-\mu) \rangle_{\mathbb{R}^n} = \theta^2 + (1-\theta)^2 - 2(1-\theta)\theta\langle\zeta,\mu\rangle = \theta^2(2 + 2\langle\zeta,\mu\rangle) - (2 + 2\langle\zeta,\mu\rangle)\theta + 1$

as a function of $\theta \in [0,1]$.

Note that: $2 + 2\langle\zeta,\mu\rangle$ is $\geq 0 (|\langle\zeta,\mu\rangle| \leq |\zeta||\mu| = 1 \Rightarrow \langle\zeta,\mu\rangle \geq -1)$

Hence $g(\theta)$ has a minimum value for $\theta = \frac{2+2\langle\zeta,\mu\rangle}{2(2+2\langle\zeta,\mu\rangle)} = \frac{1}{2}$.

This gives $g(\theta) = |\theta\zeta + (1-\theta)(-\mu)|^2 \geq g(\frac{1}{2}) = |\frac{1}{2}(\zeta - \mu)|^2$, that is our Lemma.

(if $1 + \langle\zeta,\mu\rangle = 0$, then $g(\theta) = 1 \forall \theta \in [0,1]$, hence $g(\theta) = g(\frac{1}{2})$).

Sequel to: Proof Prop 1.1 If we denote $\theta = \frac{|\xi|}{|\xi|+|\eta|}$ we see that $1 - \theta = 1 - \frac{|\xi|}{|\xi|+|\eta|} = \frac{|\eta|}{|\xi|+|\eta|}$. Because of the Lemma we derive the inequality

$$\left| \frac{|\xi|}{|\xi| + |\eta|}\zeta + \frac{|\eta|}{|\xi| + |\eta|}(-\mu) \right| \geq \frac{1}{2}|\zeta - \mu|,$$

hence

$$\frac{|\xi-\eta|}{|\xi|+|\eta|} \geq \frac{1}{2}|\zeta-\mu| = \frac{1}{2}\left|\frac{\xi}{|\xi|}-\frac{\eta}{|\eta|}\right| \tag{1.9}$$

Thus, it will be sufficient to establish the estimate,

$$|a(\infty,\xi)-a(\infty,\eta)| \leq C_1|\zeta-\mu|, \text{ that is}$$
$$|a(\infty,\zeta)-a(\infty,\mu)| \leq C_1|\zeta-\mu|, \text{ where } |\zeta|=|\mu|=1 \tag{1.10}$$

which is a Lipschitz condition verified by $a(\infty,\xi)$ on the unit ball: $\{\xi,\xi\in\mathbb{R}^n,|\xi|=1\}$.

That this is effectively so, will be established in next

Lemma 1.2 *There exists $\gamma>0$ such that*

$$|a(\infty,\xi)-a(\infty,\eta)| \leq \gamma|\xi-\eta| \; for \; |\xi|=|\eta|=1. \tag{1.11}$$

Proof Lemma 1.2 If the result is false, then, $\forall p\in\mathbb{N}$, there are vectors ξ_p,η_p of unit norm in $\mathbb{R}^n$, in such a way that the lower bound

$$|a(\infty,\xi_p)-a(\infty,\eta_p)| > p|\xi_p-\eta_p| \text{ holds, } \forall p\in\mathbb{N} \tag{1.12}$$

(hence $\xi_p\neq\eta_p\forall p\in\mathbb{N}$).

Applying twice the Bolzano–Weierstrass theorem to sequences $(\xi_p)_1^\infty,(\eta_p)_1^\infty$ we can find a common subsequence, $(\xi'_q)_1^\infty,(\eta'_q)_1^\infty$, with property that $\xi'_q\to\xi_0,\eta'_q\to\eta_0$, where, obviously, $|\xi_0|=|\eta_0|=1$.

Let us show that, in fact, $\xi_0=\eta_0$.

We use previous inequalities: $|\xi'_q-\eta'_q|\leq\frac{1}{p_{\ell_q}}|a(\infty,\xi'_q)-a(\infty,\eta'_q)|$ (derived from (1.12))

We have $p_{\ell_1}\geq\ell_1\geq 1, p_{\ell_2}\geq\ell_2\geq 2$, etc., $p_{\ell_q}\geq q$ hence

$$|\xi'_q-\eta'_q| \leq \frac{1}{q}|a(\infty,\xi'_q)-a(\infty,\eta'_q)|, \; \forall q=1,2,\ldots \tag{1.13}$$

Furthermore, we note that $a(\infty,\xi)$ as a continuous function on the unit ball in $\mathbb{R}^n$ is bounded there: $\sup\limits_{|\xi|=1}|a(\infty,\xi)|=C<\infty$.

It follows that $|\xi'_q-\eta'_q|\to 0$ as $q\to\infty$, hence $\xi_0=\eta_0$.

Let us take now a ball B with center in ξ_0: $\{\xi,|\xi-\xi_0|\leq\frac{1}{2}\}$.

Then $\xi = \xi_0 + (\xi - \xi_0)$ gives $|\xi| \geq |\xi_0| - |\xi - \xi_0| \geq \frac{1}{2}$; hence, in B the function $a(\infty, \xi)$ is in C^1. Furthermore, B is convex, so that, $\forall \xi_1, \xi_2 \in B$ we can write the mean-value formula:

$a(\infty, \xi_1) - a(\infty, \xi_2) = \langle(\xi_1 - \xi_2), \text{grad } a(\infty, \eta)\rangle$ where $\eta \in [\xi_1, \xi_2]$ – the segment which joins ξ_1 with ξ_2, and $\eta \in B$ too.

Now, as $\xi'_q \to \xi_0, \eta'_q \to \xi_0$ we find, for sufficiently large q, that ξ'_q and η'_q belong to B. Therefore we obtain

$|a(\infty, \xi'_q) - a(\infty, \eta'_q)| = |\langle \xi'_q - \eta'_q \text{ grad } a(\infty, z_q)\rangle|$ where $z_q \in [\xi'_p, \eta'_q]$ and

$$|a(\infty, \xi'_q) - a(\infty, \eta'_q)| \leq |\xi'_q - \eta'_q| \sup_{z \in B} |\text{ grad } a(\infty, z)| = \Gamma |\xi'_q - \eta'_q| \tag{1.14}$$

(due to boundedness of grad $a(\infty, z)$ on the closed ball B).

Putting together the inequalities (1.13)–(1.14) we obtain that

$$q|\xi'_q - \eta'_q| \leq |a(\infty, \xi'_q) - a(\infty, \eta'_q)| \leq \Gamma|\xi'_q - \eta'_q|, \ \forall q \in \mathbb{N} \tag{1.15}$$

; we find a contradiction when $q > \Gamma$ (remember that $\xi'_q \neq \eta'_q \forall q \in \mathbb{N}$).

Therefore, Lemma 1.2 is also established, and then we get (using Lemma 1.1)

$$|a(\infty, \xi) - a(\infty, \eta)| \leq 2\gamma|\xi - \eta|(|\xi| + |\eta|)^{-1}, \ \forall \xi, \eta \in \mathbb{R}^n/\{o\}.$$

■

Next we establish

Proposition 1.2 *The partial Fourier transform* $\tilde{a}'(\lambda, \xi) = (2\pi)^{-n/2} \int_{\mathbb{R}^n} e^{-i<x,\lambda>} a'(x, \xi) dx$ *satisfies the sequence of estimates:*

$$(1 + |\lambda|^2)^p |\tilde{a}'(\lambda, \xi)| \leq C_p, \forall \lambda \in \mathbb{R}^n, \xi \in \mathbb{R}^n/\{o\}, p = 0, 1, 2, \ldots \tag{1.16}$$

Proof Let us note first, that $\forall \xi \in \mathbb{R}^n/\{o\}$, the function $x \in \mathbb{R}^n \to a'(x, \xi)$ verifies the sequence of estimates (independent of ξ)

$$(1 + |x|^2)^p |\partial_x^\alpha a'(x, \xi)| \leq C_{p,\alpha}, \ \forall x \in \mathbb{R}^n, \forall \alpha = (\alpha_1 \ldots \alpha_n) \in \underline{\mathbb{N}}^n, p = 0, 1, 2, \ldots \tag{1.17}$$

(in fact, we assume this estimate when $|\xi| = 1$; however, we easily see that $(\partial_x^\alpha a')(x, t\xi) = (\partial_x^\alpha a')(x, \xi) \forall t > 0, x \in \mathbb{R}^n, \xi \in \mathbb{R}^n/\{o\}$; hence $(\partial_x^\alpha a')(x, \xi) = (\partial_x^\alpha a')(x, \frac{\xi}{|\xi|}), \forall \xi \neq 0$ and then

$(1+|x|^2)^p|(\partial_x^\alpha a')(x,\xi)| = (1+|x|^2)^p|(\partial_x^\alpha a')(x,\frac{\xi}{|\xi|})| \leq C_{p,\alpha}$ by assumption).

This means that the function $x \to a'(x,\xi)$ belongs to $\mathcal{S}(\mathbb{R}_x^n), \forall \xi \in \mathbb{R}^n\{o\}$, uniformly with respect to ξ (in the sense explained above).

This in turn implies that, $\forall \xi \in \mathbb{R}^n/\{o\}$, the above defined partial Fourier transform $\tilde{a}'(\lambda,\xi)$ is well defined and belongs to $\mathcal{S}(\mathbb{R}_\lambda^n)$.

Therefore, we could derive the estimates

$$(1+|\lambda|^2)^p|\tilde{a}'(\lambda,\xi)| \leq C_{p,\xi}, \text{ where } \lambda \in \mathbb{R}^n, p = 0,1,2,\ldots$$

from well-known facts of Fourier transform in $\mathcal{S}(\mathbb{R}^n)$.

However, in order to obtain the estimates in Prop. 1.2 (which are uniform with respect to ξ too) we need to repeat somehow the proof given for the Fourier transform in $\mathcal{S}(\mathbb{R}^n)$.

Thus, we note the immediate formula

$$(1+|\lambda|^2)^p\tilde{a}'(\lambda,\xi) = (2\pi)^{-n/2}\int_{\mathbb{R}^n} e^{-i<x,\lambda>}(I-\Delta_x)^p a'(x,\xi)dx \qquad (1.18)$$

where $\Delta_x = \sum_{i=1}^{n} \frac{\partial^2}{\partial x_i^2}, \lambda \in \mathbb{R}^n, \xi \in \mathbb{R}^n/\{o\}$

(which can be proved by iterated partial integration in the integral – see Prop. 4.1 in [9] Ch. 4).

Then, take q large so that: $\int_{\mathbb{R}^n} \frac{dx}{(1+|x|^2)^q} < \infty$; we then have

$(1+|\lambda|^2)^p\tilde{a}'(\lambda,\xi) = (2\pi)^{-n/2}\int_{\mathbb{R}^n} e^{-i<x,\lambda>}(1+|x|^2)^q[(I-\Delta_x)^p a'(x,\xi)](1+|x|^2)^{-q}dx$

and it follows that

$$(1+|\lambda|^2)^p|\tilde{a}'(\lambda,\xi)| \leq (2\pi)^{-n/2}\int_{\mathbb{R}^n}(1+|x|^2)^q|(I-\Delta_x)^p a'(x,\xi)|(1+|x|^2)^{-q}dx$$

$\leq C_1$, (using also the inequality $(1+|x|^2)^q|(I-\Delta_x)^p a'(x,\xi)| \leq C_{p,q}$

which follows from the assumptions (1.17) on $a'(x,\xi)$).

■

The following result about the present class of symbols expresses itself as

Proposition 1.3 *The partial Fourier transform $\tilde{a}'(\lambda,\xi)$ verifies the sequence of estimates:*

$$(1+|\lambda|^2)^p|\tilde{a}'(\lambda,\xi)-\tilde{a}'(\lambda,\eta)| \leq C_p|\xi-\eta|(|\xi|+|\eta|)^{-1}, \text{ where } \lambda \in \mathbb{R}^n, \qquad (1.19)$$

$$\xi,\eta \in \mathbb{R}^n/\{o\}, p = 0,1,2,\ldots$$

Proof Using formula (1.18) we can write the inequality:

$$(1+|\lambda|^2)^p[\tilde{a}'(\lambda,\xi)-\tilde{a}'(\lambda,\eta)] = (2\pi)^{-n/2}\int_{\mathbb{R}^n} e^{-i<x,\lambda>}(I-\Delta_x)^p[a'(x,\xi)-a'(x,\eta)]dx$$
$$= (2\pi)^{-n/2}\int_{\mathbb{R}^n} e^{-i<x,\lambda>}(1+|x|^2)^q(I-\Delta_x)^p[a'(x,\xi)-a'(x,\eta)](1+|x|^2)^{-q}dx \tag{1.20}$$

where q is sufficiently large, and then derive the estimate

$$(1+|\lambda|^2)^p|\tilde{a}'(\lambda,\xi)-\tilde{a}'(\lambda,\eta)| \le (2\pi)^{-n/2}\int_{\mathbb{R}^n}(1+|x|^2)^{-q}|B(x,\xi)-B(x,\eta)|dx$$
$$\forall\lambda\in\mathbb{R}^n, \xi,\eta\in\mathbb{R}^n/\{o\} p=0,1,2,\ldots \tag{1.21}$$

where $B(x,\xi)=(1+|x|^2)^q(I-\Delta_x)^p a'(x,\xi)$, which is homogeneous of order 0 with respect to ξ.

Now, there is a result similar to Proposition 1.1 given here as

Lemma 1.3 *There exists a positive constant C such that*

$$|B(x,\xi)-B(x,\eta)| \le C|\xi-\eta|(|\xi|+|\eta|)^{-1},\ \forall x\in\mathbb{R}^n, \xi,\eta\in\mathbb{R}^n/\{o\} \tag{1.22}$$

Proof of Lemma We first establish the Lipschitz condition on the unit ball in $\mathbb{R}^n_\xi$, uniformly for $x\in\mathbb{R}^n$, that is the estimate

$$|B(x,\zeta)-B(x,\mu)| \le C_1|\zeta-\mu|,\ \forall\zeta,\mu\in S_1=\{\xi\in\mathbb{R}^n, |\xi|=1\} \tag{1.23}$$

and $\forall x\in\mathbb{R}^n$.

In fact, in the contrary case, we would have, $\forall p=1,2,\ldots$, a couple ξ_p,η_p on S_1 and an element $x_p\in\mathbb{R}^n$, in such a way that

$$|B(x_p,\xi_p)-B(x_p,\eta_p)| > p|\xi_p-\eta_p|. \tag{1.24}$$

Apply Bolzano–Weierstrass theorem in S_1 : $\exists(\xi_{p_j})$ – subsequence of $\xi_p, \xi_{p_j}\to\xi_0\in S_1$. Then, $\exists(\eta_{p_{j_k}})_{k=1}^\infty$ – subsequence of $(\eta_{p_j})_{j=1}^\infty$, such that $\eta_{p_{j_k}}\to\eta_0\in S_1$. We have also $\xi p_{j_k}\to\xi_0$. Moreover $p_j\ge j, j_k\ge k\Rightarrow p_{j_k}\ge j_k\ge k, \forall k=1,2,\ldots$.

Therefore we get the lower bound

$$|B(x_{p_{j_k}},\xi_{p_{j_k}})-B(x_{p_{j_k}},\eta_{p_{j_k}})| > k|\xi_{p_{j_k}}-\eta_{p_{j_k}}|,\ \forall k\in\mathbb{N}. \tag{1.25}$$

If $x_{p_{j_k}} = x'_k, \xi_{p_{j_k}} = \xi'_k, \eta_{p_{j_k}} = \eta'_k$ we get from (1.25), the upper bound

$$|\xi'_k - \eta'_k| < \frac{1}{k}|B(x'_k, \xi'_k) - B(x'_k, \eta'_k)|, \ \forall k \in \mathbb{N}. \tag{1.26}$$

Actually, the function $B(x, \xi)$ is bounded on $\mathbb{R}^n \times S_1$ (as follows from the sequence of estimates defining our symbols). We obtain accordingly

$$|\xi'_k - \eta'_k| < \frac{D}{k}, \ \forall k \in \mathbb{N};$$

hence

$$\xi_0 = \lim \xi'_k = \eta_0 = \lim \eta'_k, \ |\xi_0| = |\eta_0| = 1. \tag{1.27}$$

On the other hand, let us consider again (as in the proof of Lemma 1.2), the ball: $\{\xi \in \mathbb{R}^n, |\xi - \xi_0| \leq \frac{1}{2}\}$. Here we have $|\xi| \geq \frac{1}{2}$ and then $B(x, \xi) \in C^\infty$ in this ball, $\forall x \in \mathbb{R}^n$. As $\xi'_k \to \xi_0$ and $\eta'_k \to \xi_0$ we find that $|\xi'_k - \xi_0| < \frac{1}{2}$ and $|\eta'_k - \xi_0| < \frac{1}{2}$ for $k > \bar{k}$ and due to the convexity of the ball, the segment $[\xi'_k, \eta'_k]$ is contained in this ball.

Apply the mean value theorem (with respect to the second variable)

$$B(x'_k, \xi'_k) - B(x'_k, \eta'_k) = \langle \xi'_k - \eta'_k, (\mathrm{grad}_\xi B)(x'_k, z_k) \rangle$$

for some $z_k \in [\xi'_k, \eta'_k]$ (so that $|z_k - \xi_0| \leq \frac{1}{2}$ too).

Again, we have, from the estimates defining the symbol, that

$$|\partial_{\xi_i} B(x, \xi)| \leq C \text{ for } x \in \mathbb{R}^n, \xi \in S_1, i = 1, 2, \ldots, n. \tag{1.28}$$

Note that the function $\partial_{\xi_j} B(x, \xi)$ is positively homogeneous of order -1 with respect to ξ $\forall j = 1, 2, \ldots, n$.

Thus: $(\partial_{\xi_j} B)(x, \xi) = (\partial_{\xi_j} B)(x, \frac{\xi}{|\xi|}|\xi|) = |\xi|^{-1}(\partial_{\xi_j} B)(x, \frac{\xi}{|\xi|})$, hence

$$|(\partial_{\xi_j} B)(x, \xi)| \leq \frac{C}{|\xi|} \text{ for } \xi \in \mathbb{R}^n/\{o\}, j = 1, 2, \ldots, n. \tag{1.29}$$

This implies that $|(\partial_{\xi_j} B)(x, \xi)| \leq 2C$ for $|\xi| \geq \frac{1}{2}$, hence

$$|(\partial_{\xi_j} B)(x'_k, z_k)| \leq 2C \text{ for } k > \bar{k}. \tag{1.30}$$

Combining the above estimates one gets, for $k > \bar{k}$:

$$k|\xi'_k - \eta'_k| < |B(x'_k, \xi'_k) - B(x'_k, \eta'_k)| < 2C\sqrt{n}|\xi'_k - \eta'_k|, \tag{1.31}$$

which is impossible.

This way we established the Lipschitz condition for $B(x, \zeta)$ with respect to $\zeta \in S_1$, uniformly for $x \in \mathbb{R}^n$.

Then we write, $\forall x \in \mathbb{R}^n, \xi, \eta \in \mathbb{R}^n/\{o\}$:
$B(x,\xi) - B(x,\eta) = B(x, \frac{\xi}{|\xi|}) - B(x, \frac{\eta}{|\eta|})$, whence the estimate:

$$|B(x,\xi) - B(x,\eta)| = |B(x, \frac{\xi}{|\xi|}) - B(x, \frac{\eta}{|\eta|})| \leq C_1 |\frac{\xi}{|\xi|} - \frac{\eta}{|\eta|}| \leq$$

$$2C_1 \frac{|\xi - \eta|}{|\xi| + |\eta|}, \ \forall x \in \mathbb{R}^n, \xi, \eta \in \mathbb{R}^n/\{o\} \text{ (by (1.9))}$$

which is Lemma 1.3.

Now we give the

End of Proof of Prop. 1.3 We have, using (1.21) and (1.22):

$$(1 + |\lambda|^2)^p |\tilde{a}'(\lambda, \xi) - \tilde{a}'(\lambda, \eta)| \leq (2\pi)^{-n/2} C \frac{|\xi - \eta|}{|\xi| + |\eta|} \int_{\mathbb{R}^n} (1 + |x|^2)^{-q} dx$$

$$= C_2 \frac{|\xi - \eta|}{|\xi| + |\eta|} \text{ (for large } q).$$

■

We terminate this section on the C^∞-homogeneous symbols by pointing out the following.

Corollary 1.1 *If $a(x, \xi), \mathbb{R}^n \times \mathbb{R}^n/\{o\} \to \mathbb{C}$ is a C^∞-homogeneous symbol, the estimate*

$$|a(x,\xi) - a(x,\eta)| \leq C \frac{|\xi - \eta|}{|\xi| + |\eta|}, \ x \in \mathbb{R}^n, \xi, \eta \in \mathbb{R}^n/\{o\} \tag{1.32}$$

holds true.

Proof We write the relation

$$a(x,\xi) - a(x,\eta) = a(\infty, \xi) - a(\infty, \eta) + a'(x,\xi) - a'(x,\eta)$$

Note that if $\tilde{a}'(\lambda, \xi) = (2\pi)^{-n/2} \int_{\mathbb{R}^n} e^{-i<x,\lambda>} a'(x,\xi) dx, \forall \xi \in \mathbb{R}^n/\{o\}$

then, the inverse (partial) Fourier transform

$$(2\pi)^{-n/2}\int_{\mathbb{R}^n} e^{i<x,\lambda>}\tilde{a}'(\lambda,\xi)d\lambda, \xi \in \mathbb{R}^n/\{o\}$$

gives the initial function $a'(x,\xi)$.

($\forall\xi \in \mathbb{R}^n/\{o\}, a'(x,\xi) \in \mathcal{S}(\mathbb{R}^n_x)$, hence $\tilde{a}'(\lambda,\xi) \in \mathcal{S}(\mathbb{R}^n_\lambda)$ and the Fourier inversion formula is satisfied).

Thus we have

$$a(x,\xi) - a(x,\eta) = a(\infty,\xi) - a(\infty,\eta) + (2\pi)^{-n/2}\int_{\mathbb{R}^n} e^{i<x,\lambda>}[\tilde{a}'(\lambda,\xi) - \tilde{a}'(\lambda,\eta)]d\lambda$$

whence the estimate (using Prop. 1.1 and Prop. 1.3):

$$|a(x,\xi) - a(x,\eta)| \le |a(\infty,\xi) - a(\infty,\eta)| + (2\pi)^{-n/2}\int_{\mathbb{R}^n} |\tilde{a}'(\lambda,\xi) - \tilde{a}'(\lambda,\eta)|d\lambda$$

$$\le C(|\xi-\eta|)(|\xi| + |\eta|)^{-1} + (C'_p\int_{\mathbb{R}^n}(1+|\lambda|^2)^{-p}d\lambda)(|\xi-\eta|(|\xi|+|\eta|)^{-1})$$

$$\le C_2|\xi-\eta|(|\xi|+|\eta|)^{-1} \text{ if } p \text{ is sufficiently large}), \forall x \in \mathbb{R}^n, \xi,\eta \in \mathbb{R}^n/\{o\}.$$

■

Corollary 1.2 *The inequality*

$$(1+|x|^2)^q|a'(x,\xi) - a'(x,\eta)| \le C_q|\xi-\eta|(|\xi|+|\eta|)^{-1}, \forall x \in \mathbb{R}^n, \xi,\eta \in \mathbb{R}^n/\{o\},$$
$$q = 0,1,2,\ldots \qquad (1.33)$$

holds true.

This is in fact a particular case of Lemma 1.3 (when $p = 0$).

2. Operators

In this section we shall indicate how to associate to any symbol $a(x,\xi)$ a couple of operators $A(x,D)$ and $\mathcal{A}(x,D)$ acting from $H^s(\mathbb{R}^n)$ into itself, $\forall s \in \mathbb{R}$ (and also from $\mathcal{F}^{-1}(L^p)$ into itself), as linear continuous mappings. The shortest way to do this is by referring ourselves to Ch. IV: We put $g(x,\xi) = a'(x,\xi)$. The representation formula:

$$a'(x,\xi) = (2\pi)^{-n/2}\int_{\mathbb{R}^n} e^{i<x,\lambda>}\tilde{a}'(\lambda,\xi)d\lambda, x \in \mathbb{R}^n, \xi \in \mathbb{R}^n/\{o\} \qquad (2.1)$$

where

$$\tilde{a}'(\lambda,\xi) = (2\pi)^{-n/2}\int_{\mathbb{R}^n} e^{-i<x,\lambda>}a'(x,\xi)dx \tag{2.2}$$

holds true.

Note first that $\tilde{a}'(\lambda,\xi)$ is *continuous* (hence measurable) in $\mathbb{R}^n \times \mathbb{R}^n/\{o\}$

[if $(\lambda_p,\xi_p) \to (\lambda_0,\xi_0)$ we have $e^{-i<x,\lambda_p>}a'(x,\xi_p) \to e^{-i<x,\lambda_0>}a'(x,\xi_0)\forall x \in \mathbb{R}^n$, as $a'(x,\xi) \in C(\mathbb{R}^n \times \mathbb{R}^n/\{o\})$; we also have the estimate

$$|e^{-i<x,\lambda_p>}a'(x,\xi_p)| \le |a'(x,\xi_p)| \le C_j(1+|x|^2)^{-j},\ \forall x \in \mathbb{R}^n,\ p=1,2,\dots$$

– it follows from: $(1+|x|^2)^j|a'(x,\xi)| \le C_j$ for $|\xi|=1$ and from 0-homogeneity in ξ.

Then we apply Lebesgue theorem – on dominated convergence – for j sufficiently large, in formula (2.2)].

Next, $\tilde{a}'(\lambda,\xi)$, as a function $\lambda \in \mathbb{R}^n \to \mathbb{C}(\forall\xi \in \mathbb{R}^n/\{o\})$ is obviously continuous too, hence measurable).

Let us show that

$$\sup_{\xi\in\mathbb{R}^n/\{o\}} |\tilde{a}'(\lambda,\xi)| = k(\lambda) \text{ is a function in } L^1(\mathbb{R}^n).$$

First we note that $k(\lambda)$ is a measurable function on $\mathbb{R}^n$.

In fact, if $(\xi_j)_1^\infty$ is the sequence of points with rational coordinates in $\mathbb{R}^n/\{o\}$, we have

$$k(\lambda) = \sup_{j\in\mathbb{N}} |\tilde{a}'(\lambda,\xi_j)|. \tag{2.3}$$

[This follows from the continuity of $\tilde{a}'(\lambda,\xi)\forall\xi \in \mathbb{R}^n/\{o\}$: Let us in fact define momentarily:

$\sup_{j\in\mathbb{N}}|\tilde{a}'(\lambda,\xi_j)| = k_1(\lambda)$. We obviously have $k_1(\lambda) \le k(\lambda)$.

If, for some $\bar{\lambda} \in \mathbb{R}^n$ we would have $k_1(\bar{\lambda}) < k(\bar{\lambda})$, then, there would exist a $\xi \in \mathbb{R}^n/\{o\}$, such that:

$$|\tilde{a}'(\bar{\lambda},\xi)| > k_1(\bar{\lambda}). \tag{2.4}$$

As $(\xi_j)_1^\infty$ is dense in $\mathbb{R}^n/\{o\}$, we can find ξ_k so near to ξ as to have – from continuity – $|\,|\tilde{a}'(\bar{\lambda},\xi_k)| - |\tilde{a}'(\bar{\lambda},\xi)|\,| < \mathcal{E}$, hence

$$|\tilde{a}'(\bar{\lambda},\xi_k)| > |\tilde{a}'(\bar{\lambda},\xi)| - \mathcal{E} > k_1(\bar{\lambda}) \text{ if } 0 < \mathcal{E} < |\tilde{a}'(\bar{\lambda},\xi)| - k_1(\bar{\lambda}). \tag{2.5}$$

Then, $\sup\limits_{k\in\mathbb{N}} |\tilde{a}'(\bar{\lambda},\xi_k)| = k_1(\bar{\lambda}) > k_1(\bar{\lambda})$ too, a contradiction].

Now, all functions of $\lambda \in \mathbb{R}^n, F_j(\lambda) = |\tilde{a}'(\lambda,\xi_j)|$ are continuous; then $\sup\limits_{j\in\mathbb{N}} F_j(\lambda) = k(\lambda)$ is measurable.

Next, we note also the estimates in Prop. 1.2:

$$|\tilde{a}'(\lambda,\xi)| \leq C_p(1+|\lambda|^2)^{-p} \forall \lambda \in \mathbb{R}^n, \xi \in \mathbb{R}^n/\{o\}, p = 0,1,2,\dots \quad (2.6)$$

It results that

$$\sup_{\xi\in\mathbb{R}^n/\{o\}} |\tilde{a}'(\lambda,\xi)| = k(\lambda) \leq C_p(1+|\lambda|^2)^{-p} \forall \lambda \in \mathbb{R}^n \quad (2.7)$$

This shows: $k(\cdot) \in L^1(\mathbb{R}^n)$ and even more:

$\forall s \in \mathbb{R}$, the function

$$(1+|\lambda|^2)^{|s|/2} k(\lambda) \in L^1(\mathbb{R}^n) \quad (2.8)$$

(in fact $(1+|\lambda|^2)^{|s|/2} k(\lambda) \leq C_p(1+|\lambda|^2)^{|s|/2-p}$ which belongs to $L^1(\mathbb{R}^n)$ if p is sufficiently large).

If we apply now the results in Ch. IV, we find that the operators $\mathcal{A}(x,D), A(x,D)$ given by the relations

$$\mathcal{A}(x,D)U = \mathcal{F}^{-1}\left[(2\pi)^{-n/2}\int_{\mathbb{R}^n} \tilde{a}'(\xi-\eta,\eta)\hat{U}(\eta)d\eta\right] + a(\infty,D) \quad (2.9)$$

$$A(x,D)U = \mathcal{F}^{-1}\left[(2\pi)^{-n/2}\int_{\mathbb{R}^n} \tilde{a}'(\xi-\eta,\xi)\hat{U}(\eta)d\eta\right] + a(\infty,D) \quad (2.10)$$

are well defined for $U \in \mathcal{F}^{-1}(L^p) - p \geq 1$ – or for $U \in H^s(\mathbb{R}^n)$ and are linear continuous mappings in each of these spaces

(remember that $a(\infty,D)U = (\mathcal{F}^{-1}\mathcal{M}_{a(\infty,\xi)}\mathcal{F})U$ where the function $a(\infty,\xi) = a(\infty,\frac{\xi}{|\xi|})$ is continuous and bounded in $\mathbb{R}^n/\{o\}$).

We can remark also two different representation formulas for the restriction of the operators $\mathcal{A}(x,D)$ and $A(x,D)$ to the space $\mathcal{S}(\mathbb{R}^n)$ of C^∞ – rapidly decreasing functions. Precisely we have

$$(\mathcal{A}(x,D)u)(x) = (2\pi)^{-n/2}\int_{\mathbb{R}^n} e^{i<x,\xi>} a(x,\xi)\hat{u}(\xi)d\xi, \forall x \in \mathbb{R}^n, u \in \mathcal{S}(\mathbb{R}^n) \quad (2.11)$$

and

$$(A(x,D)u)(x) = (2\pi)^{-n/2} \int_{\mathbb{R}^n} e^{i<x,\xi>} \left((2\pi)^{-n/2} \int_{\mathbb{R}^n} \bar{e}^{i<y,\xi>} a(y,\xi)u(y)dy \right) d\xi$$

$$\forall x \in \mathbb{R}^n, u \in \mathcal{S}(\mathbb{R}^n) \qquad (2.12)$$

The first formula follows from Prop. 1.2 – Ch. V, in the case $a(\infty,\xi) \equiv 0$.
In the general case, it remains to see that

$$(a(\infty,D)u)(x) = (2\pi)^{-n/2} \int_{\mathbb{R}^n} e^{i<x,\xi>} a(\infty,\xi)\hat{u}(\xi)d\xi, u \in \mathcal{S}(\mathbb{R}^n), \qquad (2.13)$$

$$\forall x \in \mathbb{R}^n.$$

Here $|a(\infty,\xi)\hat{u}(\xi)| \le \sup |a(\infty,\xi)||\hat{u}(\xi)|$ which is an $L^1(\mathbb{R}^n_\xi)$-function.
Thus, the integral in the right-hand side represents
$\mathcal{F}^{-1}[\mathcal{M}_{a(\infty,\xi)}]\mathcal{F}u$, which is precisely the definition of $a(\infty,D)u$.
As for the second formula (2.12); consider again the case $a(\infty,\xi) \equiv 0$.
We apply Prop. 1.1 – Ch. (V) and obtain accordingly

$$A(x,D)u = \mathcal{F}^{-1}\left[(2\pi)^{-n/2} \int_{\mathbb{R}^n} e^{-i<y,\xi>} a(y,\xi)u(y)dy\right] \qquad (2.14)$$

Now, the integral in (2.14) equals

$$\int_{\mathbb{R}^n} e^{-i<y,\xi>} \left[(2\pi)^{-n/2} \int_{\mathbb{R}^n} e^{i<y,\lambda>} \tilde{a}(\lambda,\xi)d\lambda\right] u(y)dy =$$

$$\int_{\mathbb{R}^n} \left((2\pi)^{-n/2} \int_{\mathbb{R}^n} e^{-i<y,\xi-\lambda>} u(y)dy\right) \tilde{a}(\lambda,\xi)d\lambda$$

$$\int_{\mathbb{R}^n} \hat{u}(\xi-\lambda)\tilde{a}(\lambda,\xi)d\lambda = \int_{\mathbb{R}^n} \tilde{a}(\xi-\eta,\xi)\hat{u}(\eta)d\eta.$$

Here we have: $|\tilde{a}(\xi-\eta,\xi)\hat{u}(\eta)| \le k(\xi-\eta)|\hat{u}(\eta)|, k \in L^1, \hat{u} \in L^1$.
Therefore the function: $\xi \to (2\pi)^{-n/2} \int_{\mathbb{R}^n} \tilde{a}(\xi-\eta,\xi)\hat{u}(\eta)d\eta$ belongs to $L^1(\mathbb{R}^n)$ and therefore the inverse Fourier transform in (2.14) is the ordinary Fourier integral in (2.12).

In the general case: $a(\infty,\xi) \not\equiv 0$, we note that (for $u \in \mathcal{S}(\mathbb{R}^n)$):
$(2\pi)^{-n/2} \int_{\mathbb{R}^n} e^{-i<y,\xi>} a(\infty,\xi)u(y)dy = a(\infty,\xi)\hat{u}(\xi)$, so that

$$(2\pi)^{-n/2} \int_{\mathbb{R}^n} e^{i<x,\xi>} a(\infty,\xi)\hat{u}(\xi)d\xi = a(\infty,D)u.$$

We shall now establish a simple relation between operators $A(x, D)$ and $\mathcal{A}(x, D)$. It is convenient here to indicate the dependence of A and $\mathcal{A}$ on the symbol a, by writing $A(x, D) = A_a(x, D), \mathcal{A}(x, D) = \mathcal{A}_a(x, D)$. Then we have

Proposition 2.1 *Let $a(x, \xi)$ be a symbol and $\bar{a}(x, \xi)$ its complex-conjugate. Then we have the equality*

$$(A_a(x, D)u, v)_{L^2} = (u, \mathcal{A}_{\bar{a}}v)_{L^2} \tag{2.15}$$

$\forall u, v \in L^2(\mathbb{R}^n)$ – over $\mathbb{C}$.

Proof Note first that if a is a symbol here satisfies (1.1)–(1.5) in Section 1 then $\bar{a}$ has the same properties. We shall use Plancherel's formula which gives, for $u, v \in L^2(\mathbb{R}^n)$:

$$\begin{aligned}
&(A_a u, v)_{L^2} = ((A_a u)^\wedge, \hat{v})_{L^2} = (a(\infty, \xi)\hat{u}(\xi), \hat{v}(\xi))_{L^2} + \\
&\left((2\pi)^{-n/2} \int_{\mathbb{R}^n} \tilde{a}'(\xi - \eta, \xi)\hat{u}(\eta)d\eta, \hat{v}(\xi)\right)_{L^2} = \\
&\int_{\mathbb{R}^n} a(\infty, \xi)\hat{u}(\xi)\bar{\hat{v}}(\xi)d\xi + (2\pi)^{-n/2} \int\int_{\mathbb{R}^n \times \mathbb{R}^n} \tilde{a}'(\xi - \eta, \xi)\hat{u}(\eta)\bar{\hat{v}}(\xi)d\eta d\xi
\end{aligned} \tag{2.16}$$

and also

$$\begin{aligned}
&(u, \mathcal{A}_a v)_{L^2} = (\hat{u}(\xi), (\mathcal{A}_{\bar{a}} v)^\wedge(\xi))_{L^2} = (\hat{u}(\xi), \bar{a}(\infty, \xi)\hat{v}(\xi))_{L^2} \\
&+ (\hat{u}(\xi), (2\pi)^{-n/2} \int_{\mathbb{R}^n} \tilde{\bar{a}}'(\xi - \eta, \eta)\hat{v}(\eta)d\eta)_{L^2} = \int_{\mathbb{R}^n} \hat{u}(\xi).\, a(\infty, \xi)\bar{\hat{v}}(\xi)d\xi + \\
&(2\pi)^{-n/2} \int_{\mathbb{R}^n} \hat{u}(\xi)\left(\int_{\mathbb{R}^n} (\bar{\tilde{\bar{a}}}'(\xi - \eta, \eta)\hat{v}(\eta)d\eta\right)d\xi = (2\pi)^{-n/2} \int_{\mathbb{R}^n} \hat{u}(\xi) \\
&\left(\int_{\mathbb{R}^n} \bar{\tilde{\bar{a}}}'(\xi - \eta, \eta)\bar{\hat{v}}(\eta)d\eta\right)d\xi + \int_{\mathbb{R}^n} \hat{u}(\xi)a(\infty, \xi)\bar{\hat{v}}(\xi)d\xi
\end{aligned} \tag{2.17}$$

Next, let us note the formula:

$$\tilde{\bar{a}}'(\lambda, \eta) = (2\pi)^{-n/2} \int_{\mathbb{R}^n} e^{-i<x,\lambda>}\bar{a}'(x, \eta)dx$$

which implies

$$\bar{\tilde{\bar{a}}}'(\lambda, \eta) = (2\pi)^{-n/2} \int_{\mathbb{R}^n} e^{i<x,\lambda>}a'(x, \eta)dx = \tilde{a}'(-\lambda, \eta).$$

It follows that

$$(2\pi)^{-n/2}\int_{\mathbb{R}^n}\bar{\tilde{a}}'(\xi-\eta,\eta)\bar{\hat{v}}(\eta)d\eta=(2\pi)^{-n/2}\int_{\mathbb{R}^n}\tilde{a}'(\eta-\xi,\eta)\bar{\hat{v}}(\eta)d\eta$$

and, from (2.17):

$$(u,\mathcal{A}_{\bar{a}}v)_{L^2}=\int_{\mathbb{R}^n}a(\infty,\xi)\hat{u}(\xi)\bar{\hat{v}}(\xi)d\xi+(2\pi)^{-n/2}\int_{\mathbb{R}^n}\hat{u}(\xi)\Big(\int_{\mathbb{R}^n}\tilde{a}'(\eta-\xi,\eta)\bar{\hat{v}}(\eta)d\eta\Big)d\xi$$

$$=\int_{\mathbb{R}^n}a(\infty,\xi)\hat{u}(\xi)\bar{\hat{v}}(\xi)d\xi+(2\pi)^{-n/2}\int\int_{\mathbb{R}^n\times\mathbb{R}^n}\tilde{a}'(\eta-\xi,\eta)\hat{u}(\xi)\bar{\hat{v}}(\eta)d\eta d\xi \qquad (2.18)$$

In the last "double" integral we make the substitution $\xi=\eta,\eta=\xi$ and obtain (2.16).

∎

Next, we are going to prove that the difference operator $A_a-\mathcal{A}_a$ is a linear continuous mapping from $H^s(\mathbb{R}^n)$ into $H^{s+1}(\mathbb{R}^n),\forall s\in\mathbb{R}$. It amounts to the inequality

$$\|(A_a-\mathcal{A}_a)U\|_{H^{s+1}}\leq C_s\|U\|_{H^s},\ \forall s\in\mathbb{R},\forall U\in H^s \qquad (2.19)$$

which means also (see Ch. I) that the operator $A_a-\mathcal{A}_a$ has order -1. Thus, let us state the following.

Proposition 2.2 *If $a(x,\xi)$ is a (K-N) symbol, the difference operator $A_a-\mathcal{A}_a$ is a linear continuous operator, $H^s\to H^{s+1}(\forall s\in\mathbb{R})$*

Proof As follows from the definition of operators A and $\mathcal{A}$ in $H^s(\mathbb{R}^n)$, we see that the Fourier transform $\mathcal{F}[(A-\mathcal{A})U](\forall U\in H^s)$ is expressed by the integral

$$(2\pi)^{-n/2}\int_{\mathbb{R}^n}[\tilde{a}'(\xi-\eta,\xi)-\tilde{a}'(\xi-\eta,\eta)]\hat{U}(\eta)d\eta \qquad (2.20)$$

(we refer to Ch.IV .2 for a thorough discussion of these kind of integrals; the one here defines a measurable function on $\mathbb{R}^n/\{o\}$, which belongs also to $L^1_{loc}(\mathbb{R}^n)$; we denote it with $w(\xi)$).

Next one considers the expression

$$w_s(\xi)=(2\pi)^{-n/2}(1+|\xi|^2)^{s+1/2}\int_{\mathbb{R}^n}[\tilde{a}'(\xi-\eta,\xi)-\tilde{a}'(\xi-\eta,\eta)]\hat{U}(\eta)d\eta$$

which can also be written as

$$w_s(\xi) = (2\pi)^{-n/2} \int\limits_{\mathbb{R}^n} (1+|\xi|^2)^{s+1/2}(1+|\eta|^2)^{-(s+1)/2}[\tilde{a}'(\xi-\eta,\xi)$$

$$-\tilde{a}'(\xi-\eta,\eta)](1+|\eta|^2)^{s+1/2}\hat{U}(\eta)d\eta \tag{2.21}$$

If we now use Prop. 1.3 (section 1) and inequality 10.10 in [9] we obtain the estimates

$$|w_s(\xi)| \leq (2\pi)^{-n/2} \int\limits_{\mathbb{R}^n} 2^{|s+1|/2}(1+|\xi-\eta|^2)^{|s+1|/2}|\tilde{a}'(\xi-\eta,\xi)$$

$$-\tilde{a}'(\xi-\eta,\eta)|(1+|\eta|^2)^{s+1/2}|\hat{U}(\eta)|d\eta \tag{2.22}$$

and

$$|\tilde{a}'(\xi-\eta,\xi) - \tilde{a}'(\xi-\eta,\eta)| \leq C_p(1+|\xi-\eta|^2)^{-p}\frac{|\xi-\eta|}{|\xi|+|\eta|}, \forall \xi,\eta \in \mathbb{R}^n/\{o\} \tag{2.23}$$

Next note the following result: *The inequality*

$$\frac{|\xi-\eta|}{|\xi|+|\eta|} \leq C(1+|\xi-\eta|^2)^{1/2}(1+|\eta|^2)^{-1/2}, \forall \xi,\eta \in \mathbb{R}^n/\{o\} \tag{2.24}$$

holds true.

Proof We have

$$\frac{|\xi-\eta|}{|\xi|+|\eta|} \leq \frac{1+|\xi-\eta|}{1+|\eta|}, \forall \xi,\eta \in \mathbb{R}^n/\{o\} \tag{2.25}$$

[in fact $(1+|\eta|)|\xi-\eta| \leq |\xi|+|\eta|+|\xi||\xi-\eta|+|\eta||\xi-\eta|$ reduces to $|\xi-\eta| \leq |\xi|+|\eta|+|\xi||\xi-\eta|$ which is obvious].

Next, there are positive constants c, C such that

$$c \leq (1+|\zeta|)(1+|\zeta|^2)^{-1/2} \leq C \forall \zeta \in \mathbb{R}^n \quad (c=1, C=\sqrt{2}) \tag{2.26}$$

[consider in fact the obvious inequalities

$1+x^2 \leq (1+x)^2 \leq 2(1+x^2)\ \forall x \geq 0$; then it follows

$\sqrt{1+x^2} \leq 1+x \leq \sqrt{2}\sqrt{1+x^2}, \forall x \geq 0$,hence $1 \leq \frac{1+x}{\sqrt{1+x^2}} \leq \sqrt{2}, \forall x \geq 0$; if $x = |\zeta|$ where $\zeta \in \mathbb{R}^n$ we get 2.26].

Then, from (2.26) we get:

$1+|\xi-\eta| \leq \sqrt{2}(1+|\xi-\eta|^2)^{1/2}$ and $1+|\eta| \geq (1+|\eta|^2)^{1/2}$

hence $\frac{1+|\xi-\eta|}{1+|\eta|} \leq \frac{\sqrt{2}(1+|\xi-\eta|^2)^{1/2}}{(1+|\eta|^2)^{1/2}}$ and then, from (2.25) we derive

$$\frac{|\xi-\eta|}{|\xi|+|\eta|} \leq \sqrt{2}(1+|\xi-\eta|^2)^{1/2}(1+|\eta|^2)^{-1/2}, \forall \xi, \eta \in \mathbb{R}^n/\{o\}$$

which is (2.24).

Thus, we find that (using (2.23) and (2.24)):

$$\begin{aligned}&(1+|\xi-\eta|^2)^{|s+1|/2}|\tilde{a}'(\xi-\eta,\xi)-\tilde{a}'(\xi-\eta,\eta)|(1+|\eta|^2)^{s+1/2}|\hat{U}(\eta)| \quad (2.27)\\ &\leq C_p(1+|\xi-\eta|^2)^{|s+1|/2-p}(1+|\xi-\eta|^2)^{1/2}(1+|\eta|^2)^{s/2}|\hat{U}(\eta)|\\ &\qquad \text{for } \xi \in \mathbb{R}^n/\{o\}, \eta \in \mathbb{R}^n/\{o\}.\end{aligned}$$

This entails the estimate (using (2.22)):

$$|w_s(\xi)| \leq C_{s,p}\int_{\mathbb{R}^n}(1+|\xi-\eta|^2)^{|s+1|/2-p+1/2}(1+|\eta|^2)^{s/2}|\hat{U}(\eta)|d\eta \quad (2.28)$$
$$\forall \xi \in \mathbb{R}^n/\{o\}.$$

The right-hand side integral (for large p) is a convolution between the function $(1+|\lambda|^2)^{|s+1|/2+1/2-p}$ which belongs to $L^1(\mathbb{R}^n)$
and the function $(1+|\eta|^2)^{s/2}|\hat{U}(\eta)|$ which belongs to L^2 (remember that $U \in H^s(\mathbb{R}^n)$).

By Young's theorem, the right hand side integral in (2.28) belongs to $L^2(\mathbb{R}^n)$ (its L^2-norm is estimated by $C \parallel U \parallel_{H^s}$).

As $w_s(\xi)$ is a measurable function which is estimated by an L^2-function, it follows that

$$\int_{\mathbb{R}^n}|w_s(\xi)|^2 d\xi \text{ exists and is } \leq C \parallel U \parallel_{H^s}.$$

This means that the function $w(\xi) = \int_{\mathbb{R}^n}[\tilde{a}'(\xi-\eta,\xi)-\tilde{a}'(\xi-\eta,\eta)]\hat{U}(\eta)d\eta$ is measurable and that $(1+|\xi|^2)^{s+1/2}w(\xi) \in L^2(\mathbb{R}^n)$. Thus we see that $(A-\mathcal{A})U \in H^{s+1}$ and the estimate

$$\parallel (A-\mathcal{A})U \parallel_{H^{s+1}} \leq C_s \parallel U \parallel_{H^s}, \forall U \in H^s,$$

holds true.

■

Chapter VII
Compactness of operator $A(x,D) - \mathcal{A}(x,D)$ in the space $\mathcal{F}^{-1}(L^1(\mathbb{R}^n))(A(x,D) \equiv \mathcal{A}(x,D)$-modulo the compact operators)

Introduction

This section is dedicated to a result about the compactness of the difference operator $A(x,D) - \mathcal{A}(x,D)$ where the corresponding symbol is in a special class (of the kind appearing in Ch. IV) but with a few more supplementary properties. The space considered – $(\mathcal{F}^{-1}(L^1(\mathbb{R}^n)) = \mathcal{B}_{1,0}(\mathbb{R}^n)$ – has also been previously explained.

1.

Therefore we shall consider measurable functions $a(x,\xi), \mathbb{R}^n \times \mathbb{R}^n/\{o\} \to \mathbb{C}$, admitting a representation formula

$$a(x,\xi) = (2\pi)^{-n/2} \int_{\mathbb{R}^n} e^{i<x,\lambda>} \tilde{a}(\lambda,\xi) d\lambda, \ \forall (x,\xi) \in \mathbb{R}^n \times \mathbb{R}^n/\{o\} \tag{1.1}$$

where $\tilde{a}(\lambda,\xi)$ is also a measurable function, $\mathbb{R}^n \times \mathbb{R}^n/\{o\} \to \mathbb{C}$, such that: $\lambda \to \tilde{a}(\lambda,\xi)$ is measurable, $\mathbb{R}^n \to \mathbb{C}$, $\forall \xi \in \mathbb{R}^n/\{o\}$, and $|\tilde{a}(\lambda,\xi)| \leq k(\lambda)$ which belongs to $L^1(\mathbb{R}^n)$, $\forall \xi \in \mathbb{R}^n/\{o\}$.

It results (see Ch. IV -1), that the operators defined by:

$$A(x,D)U = \mathcal{F}^{-1}\big[(2\pi)^{-n/2} \int_{\mathbb{R}^n} \tilde{a}(\xi - \eta, \xi) \hat{U}(\eta) d\eta\big], \tag{1.2}$$

$$\mathcal{A}(x,D)U = \mathcal{F}^{-1}\big[(2\pi)^{-n/2} \int_{\mathbb{R}^n} \tilde{a}(\xi - \eta, \eta) \hat{U}(\eta) d\eta\big], \tag{1.3}$$

when $U \in \mathcal{F}^{-1}(L^1(\mathbb{R}^n))$ are well-defined linear continuous operators from $\mathcal{B}_{1,0}$ into itself, with operator norm $\leq (2\pi)^{-n/2} \parallel k(\cdot) \parallel_{L^1(\mathbb{R}^n)}$.

Our main goal in this section : to establish validity of the following.

Theorem *Let us assume the supplementary properties of the symbol $a(x,\xi)$:*

i) The function $(1 + |\lambda|)k(\lambda)$ belongs to $L^1(\mathbb{R}^n)$.

ii) There exists a continuous function $\varphi(t), [0,\infty) \to (0,\infty)$, such that $\lim_{t\to\infty} \varphi(t) = 0$ and the inequality

$$|\tilde{a}(\lambda,\xi) - \tilde{a}(\lambda,\tau)| \leq (1 + |\xi - \tau|)\varphi(|\xi|)k(\lambda), \ \lambda \in \mathbb{R}^n, \ \xi,\tau \in \mathbb{R}^n/\{o\} \tag{1.4}$$

holds true.

iii) The relations

$$\int_{\mathbb{R}^n} |x|\ |a(x,\xi)|dx < C,\ \forall \xi \in \mathbb{R}^n/\{o\} \tag{1.5}$$

and

$$\lim_{|h|\to 0} \int_{|\xi|\le\Lambda} \int_{\mathbb{R}^n} |a(x,\xi+h) - a(x,\xi)|dx\ d\xi = 0 \tag{1.6}$$

hold, $\forall \Lambda > 0$.

Then, the difference operator $A(x,D) - \mathcal{A}(x,D)$ *maps any bounded set in* $\mathcal{B}_{1,0}$ *into a relatively compact set in* $\mathcal{B}_{1,0}$.

Remark The existence of the integral (1.6) can be derived from the following observations:

First, the function $x \to a(x,\xi)$ is continuous in $\mathbb{R}^n \forall \xi \in \mathbb{R}^n/\{o\}$ (as a consequence of (1.1)) and also $|a(x,\xi)| \le (2\pi)^{-n/2} \parallel k \parallel_{L^1}$. This shows $\int_{|x|\le 1} |a(x,\xi)|dx$ exists, $\forall \xi \in \mathbb{R}^n/\{o\}$ and is $\le C_1$ for all ξ.

For $|x| \ge 1$ we have $|a(x,\xi)| \le |x|\ |a(x,\xi)|$ so that, using (1.5) one gets

$$\int_{|x|\ge 1} |a(x,\xi)|dx \le \int_{|x|\ge 1} |x|\ |a(x,\xi)|dx < C,\ \forall \xi \in \mathbb{R}^n/\{o\}. \tag{1.7}$$

We find a constant $C' > 0$, such that

$$\int_{\mathbb{R}^n} |a(x,\xi)|dx \le C',\ \forall \xi \in \mathbb{R}^n/\{o\} \tag{1.8}$$

which shows also that

$$\int_{\mathbb{R}^n} |a(x,\xi+h) - a(x,\xi)|dx \le 2C',\ \forall \xi \in \mathbb{R}^n/\{o\}, \xi \ne -h$$

Let us establish also the measurability of the function, $\mathbb{R}^n/\{o,-h\} \to R$

$$\varphi(\xi) = \int_{\mathbb{R}^n} |a(x,\xi+h) - a(x,\xi)|dx$$

Note that the function $(x,\xi) \in \mathbb{R}^n \times \mathbb{R}^n/\{o\} \to \mathbb{C},\ a(x,\xi)$, is measurable and bounded; this shows that the function $(x,\xi) \in \mathbb{R}^{\times}\mathbb{R}^n/\{-h\} \to \mathbb{C}, a(x,\xi+h)$, has same property; hence:

$$\int\int_{\substack{|\xi|\le R\\ |x|\le n}} |a(x,\xi+h)-a(x,\xi)|dxd\xi \text{ exists, } \forall R>0, n=1,2,\dots$$

Then, $\forall n \in \mathbb{N}$, the function $\varphi_n(\xi) = \int_{|x|\le n} |a(x,\xi+h)-a(x,\xi)|dx$ is measurable in $\mathbb{R}^n/\{o,-h\}$; consequently, the function $\varphi(\xi) = \lim_{n\to\infty} \varphi_n(\xi)$ is also measurable.

2. The proof

We start with a criterion of relative compactness for sets in $\mathcal{B}_{1,0}(\mathbb{R}^n)$.

We note the obvious fact: a set $\mathcal{U} = \{U\}$ in $\mathcal{B}_{1,0}$ is relatively compact ("each sequence $\{U_n\}$ in $\mathcal{U}$ has a convergent subsequence") if and only if the set of Fourier transforms $\{\mathcal{F}(U)\}$ is relatively compact in $L^1(\mathbb{R}^n)$. Furthermore (a well-known classical result, Th. 21p. 301 [3]), the following 3 conditions are sufficient for the relative compactness of a set $\mathcal{L}$ in $L^1(\mathbb{R}^n)$:

a) $\int_{\mathbb{R}^n} |f(\xi)|d\xi \le C$, $\forall f \in \mathcal{L}$

b) $\lim_{\Lambda\to\infty} \int_{|\xi|\ge\Lambda} |f(\xi)|d\xi = 0$ uniformly on $\mathcal{L}$

c) $\lim_{|\tau|\to 0} \int_{\mathbb{R}^n} |f(\xi+\tau)-f(\xi)|d\xi = 0$, uniformly on $\mathcal{L}$.

We can now state

Proposition 2.1 *Let $\mathcal{K}$ be a set in $\mathcal{B}_{1,0}$, satisfying conditions:*

α) If $\Lambda \in \mathbb{R}, \Lambda > 0$, then

$$\lim_{|h|\to 0} \int_{|\xi|\le\Lambda} |(\mathcal{F}U)(\xi+h) - \mathcal{F}(U)(\xi)|d\xi = 0 \tag{2.1}$$

uniformly for $U \in \mathcal{K}$.

β) There exists a continuous function $\psi(t), [0,\infty) \to (0,\infty)$, such that $\psi(t) \ge \gamma > 0$ $\forall t \ge 0, \lim_{t\to\infty} \psi(t) = \infty$ and

$$\int_{\mathbb{R}^n} \psi(|\xi|)|(\mathcal{F}U)(\xi)|d\xi \le C, \ \forall U \in \mathcal{K} \tag{2.2}$$

It results: $\mathcal{K}$ is a relatively compact set in $\mathcal{B}_{1,0}$.

Proof of Prop. 2.1 It will be shown that the set $\{\mathcal{F}U\}_{U\in\mathcal{K}}$ satisfies conditions a)-b)-c) above.

We first use condition β); we find $\gamma \int_{\mathbb{R}^n} |(\mathcal{F}U)(\xi)|d\xi \le C$ $\forall U \in \mathcal{K}$ which is a).

Next: still from β) we derive, $\forall M > 0$, that $\psi(t) \geq M$ if $t \geq t_M$; then $\int_{|\xi|\geq t_M} \psi(|\xi|)|(\mathcal{F}U)(\xi)|d\xi \leq C$ implies that

$$M \int_{|\xi|\geq t_M} |(\mathcal{F}U)(\xi)|d\xi \leq C, \int_{|\xi|\geq t_M} |(\mathcal{F}U)(\xi)|d\xi \leq \frac{C}{M}, \forall U \in \mathcal{K}$$

Thus, $\forall \varepsilon > 0$, if $\frac{C}{M} \leq \varepsilon$ we get $\int_{|\xi|\geq\Lambda} |(\mathcal{F}U)(\xi)|d\xi \leq \varepsilon$ for $\Lambda > t_{M(\varepsilon)}$ and $\forall U \in \mathcal{K}$, which is b).

Finally, we show that

$$\lim_{\tau\to 0} \int_{\mathbb{R}^n} |(\mathcal{F}U)(\xi+\tau) - (\mathcal{F}U)(\xi)|d\xi = 0, \text{ uniformly for } U \in \mathcal{K}$$

(condition c) for the set $(\mathcal{F}U)_{U\in\mathcal{K}}$).

We write the obvious decomposition:

$$\int_{\mathbb{R}^n} |(\mathcal{F}U)(\xi+\tau) - (\mathcal{F}U)(\xi)|d\xi = \int_{|\xi|\leq\Lambda+1} |(\mathcal{F}U)(\xi+\tau) - (\mathcal{F}U)(\xi)|d\xi +$$

$$\int_{|\xi|\geq\Lambda+1} |(\mathcal{F}U)(\xi+\tau) - (\mathcal{F}U)(\xi)|d\xi \quad \text{where } \Lambda > 0. \tag{2.3}$$

Consider the set in $\mathbb{R}^n$: $\{\xi \in \mathbb{R}^n, |\xi+\tau| \geq \Lambda\}$ (where $\tau \in \mathbb{R}^n$ is fixed); (it is the exterior of the ball of center $-\tau$ and radius Λ). Let now: $\xi \in \mathbb{R}^n$, $|\xi| \geq \Lambda + 1$ and $\tau \in \mathbb{R}^n$, $|\tau| \leq 1$. It follows that $|\xi+\tau| \geq |\xi| - |\tau| \geq \Lambda$. Thus we get:

$$\{\xi \in \mathbb{R}^n, |\xi| \geq \Lambda+1\} \subset \{\xi \in \mathbb{R}^n, |\xi+\tau| \geq \Lambda\} \text{ if } |\tau| \leq 1 \tag{2.4}$$

Accordingly we find, for $|\tau| \leq 1$, that

$$\int_{|\xi|\geq\Lambda+1} |(\mathcal{F}U)(\xi+\tau) - (\mathcal{F}U)(\xi)|d\xi \leq \int_{|\xi+\tau|\geq\Lambda} |(\mathcal{F}U)(\xi+\tau)|d\xi + \int_{|\xi|\geq\Lambda+1} |(\mathcal{F}U)(\xi)|d\xi$$

$$= \int_{|\eta|\geq\Lambda} |(\mathcal{F}U)(\eta)|d\eta + \int_{|\eta|\geq\Lambda+1} |(\mathcal{F}U)(\eta)|d\eta \leq 2 \int_{|\eta|\geq\Lambda} |(\mathcal{F}U)(\eta)|d\eta \tag{2.5}$$

Take now $\varepsilon > 0$. As seen above, we can find Λ_ε such that
$2\int_{|\eta|\geq\Lambda_\varepsilon} |(\mathcal{F}U)(\eta)|d\eta < \frac{\varepsilon}{2}$ for all $U \in \mathcal{K}$. (condition b) for $\{\mathcal{F}U\}_{U\in\mathcal{K}}$)
Next, using assumption α) we find, for $|\tau| < \delta(\varepsilon)$, that

$$\int_{|\xi|\leq\Lambda+1} |(\mathcal{F}U)(\xi+\tau) - (\mathcal{F}U)(\xi)|d\xi < \frac{\varepsilon}{2}, \forall U \in \mathcal{K}.$$

The Prop. 2.1 is therefore established.

Proof of Theorem - continuation Let $T = A(x,D) - \mathcal{A}(x,D)$ and Ω be a bounded set in $\mathcal{B}_{1,0}$ (therefore: $\int_{\mathbb{R}^n} |\mathcal{F}u(\xi)|d\xi \le C,\ \forall u \in \Omega$). We shall see that the image set $\mathcal{K} = \{T(u), u \in \Omega\}$ satisfies conditions $\alpha) - \beta)$ in Prop. 2.1.

Remember that condition $\alpha)$ reads as follows:

$$\lim_{|h|\to 0} \int_{|\xi|\le\Lambda} |\mathcal{F}(Au - \mathcal{A}u)(\xi + h) - \mathcal{F}(Au - \mathcal{A}u)(\xi)|d\xi = 0$$

$$\text{uniformly for } \int_{\mathbb{R}^n} |(\mathcal{F}u)(\xi)|d\xi \le C$$

It suffices therefore to establish separately that

$$\lim_{|h|\to 0} \int_{|\xi|\le\Lambda} |\mathcal{F}(Au)(\xi + h) - \mathcal{F}(Au)(\xi)|d\xi = 0 \tag{2.6}$$

and

$$\lim_{|h|\to 0} \int_{|\xi|\le\Lambda} |\mathcal{F}(\mathcal{A}u)(\xi + h) - \mathcal{F}(\mathcal{A}u)(\xi)|d\xi = 0 \tag{2.7}$$

$$\text{for all } \Lambda > 0, \text{ uniformly if } \int_{\mathbb{R}^n} |(\mathcal{F}u)(\xi)|d\xi \le C$$

We have obviously that

$$(\mathcal{F}(A(x,D))u)(\xi) = (2\pi)^{-n/2} \int_{\mathbb{R}^n} \tilde{a}(\xi - \eta, \xi)\hat{u}(\eta)d\eta \tag{2.8}$$

and

$$(\mathcal{F}(\mathcal{A}(x,D))u)(\xi) = (2\pi)^{-n/2} \int_{\mathbb{R}^n} \tilde{a}(\xi - \eta, \eta)\hat{u}(\eta)d\eta \tag{2.9}$$

$$\text{for } u \in \mathcal{F}^{-1}(L^1(\mathbb{R}^n)).$$

Accordingly we find that

$$\mathcal{F}(Au)(\xi + h) - \mathcal{F}(Au)(\xi) = (2\pi)^{-n/2} \int_{\mathbb{R}^n} [\tilde{a}(\xi + h - \eta, \xi + h) - \tilde{a}(\xi - \eta, \xi)]\tilde{u}(\eta)d\eta \tag{2.10}$$

and also that

$$\mathcal{F}(\mathcal{A}u)(\xi + h) - \mathcal{F}(\mathcal{A}u)(\xi) = (2\pi)^{-n/2} \int_{\mathbb{R}^n} [\tilde{a}(\xi + h - \eta, \eta) - \tilde{a}(\xi - \eta, \eta)]\tilde{u}(\eta)d\eta \tag{2.11}$$

for $u \in \mathcal{F}^{-1}(L^1(\mathbb{R}^n))$.

Let us note the obvious identity:

$$\tilde{a}(\xi+h-\eta,\xi+h)-\tilde{a}(\xi-\eta,\xi)=\tilde{a}(\xi+h-\eta,\xi+h)-\tilde{a}(\xi-\eta,\xi+h)+\tilde{a}(\xi-\eta,\xi+h)-\tilde{a}(\xi-\eta,\xi). \tag{2.12}$$

Note also that, from (1.1) the symbol $a(x,\xi), \forall\xi\in\mathbb{R}^n/\{o\}$ is the inverse partial Fourier transform of the integrable function $\lambda\to\tilde{a}(\lambda,\xi)$.

Hence, $a(x,\xi)=\mathcal{F}_\lambda^{-1}[\tilde{a}(\lambda,\xi)]$ in $\mathcal{S}'(\mathbb{R}^n)$-sense and $\tilde{a}(\lambda,\xi)=\mathcal{F}_x[a(x,\xi)]$ in $\mathcal{S}'(\mathbb{R}^n)$-sense. On the other hand, from (1.8) we derive that $a(x,\xi)\in L^1(\mathbb{R}^n_x)\forall\xi\in\mathbb{R}^n/\{o\}$ and accordingly we obtain that

$$\tilde{a}(\lambda,\xi)=(2\pi)^{-n/2}\int_{\mathbb{R}^n}e^{-i<x,\lambda>}a(x,\xi)dx,\ \forall\lambda\in\mathbb{R}^n,\ \forall\xi\in\mathbb{R}^n/\{o\} \tag{2.13}$$

in ordinary sense.

This entails the following representation formula:

$$\tilde{a}(\xi+h-\eta,\xi+h)-\tilde{a}(\xi-\eta,\xi+h)=(2\pi)^{-n/2}\int_{\mathbb{R}^n}e^{-i<x,\xi+h-\eta>}a(x,\xi+h)dx-$$
$$(2\pi)^{-n/2}\int_{\mathbb{R}^n}e^{-i<x,\xi-\eta>}a(x,\xi+h)dx=(2\pi)^{-n/2}\int_{\mathbb{R}^n}e^{-i<x,\xi-\eta>}[e^{-i<x,h>}-1]a(x,\xi+h)dx. \tag{2.14}$$

Note now the following elementary estimate: $|e^{-i<x,h>}-1|\le|x|\,|h|$ (in fact $|e^{-i<x,h>}-1|=|\cos<x,h>-1-i\sin<x,h>|=[(\cos<x,h>-1)^2+\sin^2<x,h>]^{1/2}=(2-2\cos<x,h>)^{1/2}=\sqrt{2}\left(2\sin^2\frac{<x,h>}{2}\right)^{1/2}=2\left|\sin\frac{<x,h>}{2}\right|$; also $|\sin\alpha|\le|\alpha|\ \forall\alpha\in\mathbb{R}$ –

[for, if $\alpha>0$, consider $g(\alpha)=\sin\alpha-\alpha$; $g(0)=0$; $g'(\alpha)=\cos\alpha-1\le0$, hence $\sin\alpha\le\alpha$ for $\alpha>0$. Now, if $0<\alpha\le1$, we have that $\sin\alpha>0$, hence $|\sin\alpha|=\sin\alpha<\alpha$; if $1<\alpha$ we have $|\sin\alpha|\le1<\alpha=|\alpha|$; thus $|\sin\alpha|\le|\alpha|=\alpha\ \forall\alpha\ge0$; if $\alpha<0$ we have $\sin\alpha=-\sin(-\alpha)$; $|\sin\alpha|=|\sin(-\alpha)|\le|-\alpha|=|\alpha|$].

Altogether this gives the inequality: $|e^{-i<x,h>}-1|\le2\frac{|<x,h>|}{2}\le|x|\,|h|$. We derive therefore the estimate (using 2.14)) and assumption iii) of Theorem:

$$|\tilde{a}(\xi+h-\eta,\xi+h)-\tilde{a}(\xi-\eta,\xi+h)|\le(2\pi)^{-n/2}|h|\int_{\mathbb{R}^n}|x|\,|a(x,\xi+h)|dx\le C|h| \tag{2.15}$$

Next, we need a (similar) estimate for the expression: $\tilde{a}(\xi-\eta,\xi+h)-\tilde{a}(\xi-\eta,\xi)$ which is (obviously) given by the representation formula:

$$(2\pi)^{-n/2}\int_{\mathbb{R}^n} e^{-i<x,\xi-\eta>}a(x,\xi+h)dx-(2\pi)^{-n/2}\int_{\mathbb{R}^n} e^{-i<x,\xi-\eta>}.\,a(x,\xi)dx$$
$$=(2\pi)^{-n/2}\int_{\mathbb{R}^n} e^{-i<x,\xi-\eta>}[a(x,\xi+h)-a(x,\xi)]dx. \tag{2.16}$$

We obtain accordingly

$$|\tilde{a}(\xi-\eta,\xi+h)-\tilde{a}(\xi-\eta,\xi)|\leq(2\pi)^{-n/2}\int_{\mathbb{R}^n}|a(x,\xi+h)-a(x,\xi)|dx,\ \forall\xi\in\mathbb{R}^n/\{o\},\eta\in\mathbb{R}^n \tag{2.17}$$

We are thus ready to evaluate the integral in (2.6), that is
$\int_{|\xi|\leq\Lambda}|\mathcal{F}(Au)(\xi+h)-\mathcal{F}(Au)(\xi)|dx$; first we have, by (2.10) that

$$|\mathcal{F}(Au)(\xi+h)-\mathcal{F}(Au)(\xi)|\leq(2\pi)^{-n/2}\int_{\mathbb{R}^n}|\tilde{a}(\xi+h-\eta,\xi+h)$$
$$-\tilde{a}(\xi-\eta,\xi)|\,|\hat{u}(\eta)|d\eta\leq(2\pi)^{-n/2}\int_{\mathbb{R}^n}|\tilde{a}(\xi+h-\eta,\xi+h)-\tilde{a}(\xi-\eta,\xi+h)|\,|\hat{u}(\eta)|d\eta$$
$$+\ (2\pi)^{-n/2}\int_{\mathbb{R}^n}|\tilde{a}(\xi-\eta,\xi+h)-\tilde{a}(\xi-\eta,\xi)|\,|\hat{u}(\eta)|d\eta$$

using (2.15) and (2.17), this is

$$\leq C_1|h|\int_{\mathbb{R}^n}|\hat{u}(\eta)|d\eta+C_2\int_{\mathbb{R}^n_\eta}\int_{\mathbb{R}^n_x}|a(x,\xi+h)-a(x,\xi)||\hat{u}(\eta)|dxd\eta\leq$$
$$C_3|h|+C\int_{\mathbb{R}^n_x}|a(x,\xi+h)-a(x,\xi)|dx$$

for u in a bounded set of $\mathcal{B}_{1,0}$

– meaning that $\int_{\mathbb{R}^n_\eta}|\hat{u}(\eta)|d\eta\leq\Gamma$ for those u–.

Therefore, we obtain that:

$$\int\limits_{|\xi|\leq\Lambda}|\mathcal{F}(Au)(\xi+h)-\mathcal{F}(Au)(\xi)|d\xi\leq C_3|h|\int\limits_{|\xi|\leq\Lambda}d\xi+C_4\int_{|\xi|\leq\Lambda}\Big(\int\limits_{\mathbb{R}^n_x}|a(x,\xi+h)$$
$$-a(x,\xi)|dx\Big)d\xi \tag{2.18}$$

for u in a bounded set of $\mathcal{B}_{1,0}$.

The terms in the right-hand side $\to 0$ as $|h|\to 0$ (see assumption iii)).

Next, we shall prove also that

$\lim_{|h|\to 0} \int_{|\xi|\leq\Lambda} |\mathcal{F}(\mathcal{A}u)(\xi+h) - \mathcal{F}(\mathcal{A}u)(\xi)|d\xi = 0$, uniformly for u in a bounded set of $\mathcal{B}_{1,0}$

$\forall\Lambda > 0$ (condition (2.7) above).

(the proof appears shorter than the previously given reasonings).

We have (from (2.11))

$$|\mathcal{F}(\mathcal{A}u)(\xi+h) - \mathcal{F}(\mathcal{A}u)(\xi)| \leq (2\pi)^{-n/2} \int_{\mathbb{R}^n} |\tilde{a}(\xi+h-\eta,\eta) - \tilde{a}(\xi-\eta,\eta)|\, |\tilde{u}(\eta)|d\eta \tag{2.19}$$

If we use (2.13) we obtain

$$\tilde{a}(\xi+h-\eta,\eta) = (2\pi)^{-n/2} \int_{\mathbb{R}^n} e^{-i<x,\xi+h-\eta>} a(x,\eta)dx \tag{2.20}$$

and

$$\tilde{a}(\xi-\eta,\eta) = (2\pi)^{-n/2} \int_{\mathbb{R}^n} e^{-i<x,\xi-\eta>} a(x,\eta)dx, \tag{2.21}$$

; therefore one gets

$$\tilde{a}(\xi+h-\eta,\eta)-\tilde{a}(\xi-\eta,\eta) = (2\pi)^{-n/2} \int_{\mathbb{R}^n} e^{-i<x,\xi-\eta>}[e^{-i<x,h>}-1]a(x,\eta)dx \tag{2.22}$$

and accordingly the inequality:

$$|\tilde{a}(\xi+h-\eta,\eta) - \tilde{a}(\xi-\eta,\eta)| \leq C \int_{\mathbb{R}^n} |e^{-i<x,h>} - 1|\, |a(x,\eta)|dx \leq$$

$$C_1 \int_{\mathbb{R}^n} |h|\, |x|\, |a(x,\eta)|dx \leq C_2|h|. \tag{2.23}$$

(we use again assumption iii)).

It follows that, using (2.19) and (2.23):

$$|\mathcal{F}(\mathcal{A}u)(\xi+h) - \mathcal{F}(\mathcal{A}u)(\xi)| = C \int_{\mathbb{R}^n} |h|\, |\hat{u}(\eta)|d\eta = C''|h| \tag{2.24}$$

for u in a bounded set of $\mathcal{F}^{-1}(L^1)$.

Therefore:

$\int_{|\xi|\leq\Lambda} |\mathcal{F}(\mathcal{A}u)(\xi+h) - \mathcal{F}(\mathcal{A}u)(\xi)|d\xi \leq (C'' \int_{|\xi|\leq\Lambda} d\xi)|h| \to 0$ if $|h| \to 0 \forall\Lambda > 0$.

Thus, we ended the proof of the fact that the set $\{Tu\}_{u\in\Omega}$ satisfies condition α in Prop. 2.1.

Proof of Theorem, end

In this last part of the proof we show that, if Ω is a bounded set in $\mathcal{B}_{1,0}$, the range of the operator $T = A - \mathcal{A} : K = \{Tu, u \in \Omega\}$ satisfies condition β) in Prop. 2.1. Precisely, we establish that, with $\psi(t) = \frac{1}{\varphi(t)}, t \geq 0$, where $\varphi(\cdot)$ appears in the Assumption ii) of the Theorem, we have an estimate of the form

$$\int_{\mathbb{R}^n} \psi(|\xi|)|(\mathcal{F}U)(\xi)|d\xi \leq C, \forall U \in \mathcal{K}. \tag{3.1}$$

As $\varphi(t) > 0 \forall t \geq 0$, we first see that the function $\psi = \frac{1}{\varphi}$ is well-defined on $[0, \infty)$, and, furthermore: $\psi(t) > 0 \ \forall t \geq 0$. Also, as $\varphi(t) \to 0$ for $t \to +\infty$, it follows that $\psi(t) \to +\infty$ as $t \to +\infty$, whence $\psi(t) \geq 1$ for $t \geq t_1 > 0$. If $\gamma_1 = \inf_{0 \leq t \leq t_1} \psi(t)$ we see, using continuity of φ and ψ, that $\gamma_1 > 0$. Let us put now: $\gamma = \inf(1, \gamma_1)$; then $\gamma > 0$ and $\psi(t) \geq \gamma \forall t \geq 0$.

Therefore, it remains to prove that

$$\int_{\mathbb{R}^n} \psi(|\xi|)| \ |(\mathcal{F}(A - \mathcal{A})u)(\xi)|d\xi \leq C \ \forall u \in \Omega \tag{3.2}$$

Actually, the term under the integral sign is nothing else than $\psi(|\xi|)(2\pi)^{-n/2} \int_{\mathbb{R}^n} [\tilde{a}(\xi - \eta, \xi) - \tilde{a}(\xi - \eta, \eta)]\hat{u}(\eta)d\eta$; its absolute value becomes, using assumption ii) of the Theorem,

$$\leq \psi(|\xi|)(2\pi)^{-n/2} \int_{\mathbb{R}^n} (1 + |\xi - \eta|)\varphi(|\xi|)k(\xi - \eta)|\hat{u}(\eta)|d\eta =$$

$$(2\pi)^{-n/2} \int_{\mathbb{R}^n} (1 + |\xi - \eta|)k(\xi - \eta)|\hat{u}(\eta)|d\eta \tag{3.3}$$

It follows readily that the estimate

$$\begin{aligned}
\int_{\mathbb{R}^n} \psi(|\xi|)|(\mathcal{F}(A - \mathcal{A})u)(\xi)|d\xi &\leq \int_{\mathbb{R}^n} ((2\pi)^{-n/2} \int_{\mathbb{R}^n} (1 + |\xi - \eta|)k(\xi - \eta)|\hat{u}(\eta)|d\eta)d\xi \\
&= (2\pi)^{-n/2} \int_{\mathbb{R}^n} (\int_{\mathbb{R}^n} (1 + |\xi - \eta|)k(\xi - \eta)d\xi)|\hat{u}(\eta)|d\eta \\
&= (2\pi)^{-n/2} (\int_{\mathbb{R}^n} (1 + |\lambda|)k(\lambda)d\lambda) \int_{\mathbb{R}^n} |\hat{u}(\eta)|d\eta \\
&\leq C_2 \text{ for } u \in \Omega \text{ , is verified}
\end{aligned} \tag{3.4}$$

(using assumption i) of the Theorem).

This ends the proof.

Remark An important instance of symbols $a(x,\xi)$ which satisfy all conditions i)–ii)–iii) of the Theorem is furnished by the (Kohn-Nirenberg) homogeneous and C^∞-symbols which were explained in Ch. VI. Presently, we shall also assume that: $a(\infty,\xi) = 0 \forall \xi \in \mathbb{R}^n/\{o\}$, so that $a(x,\xi) = a'(x,\xi)$ (this is not a real restriction when studying operator $A - \mathcal{A}$!). Then $\tilde{a}(\lambda,\xi)$ will be the partial Fourier transform (Prop. 1.2 – Ch.VI):

$$\tilde{a}(\lambda,\xi) = (2\pi)^{-n/2} \int_{\mathbb{R}^n} e^{-i<x,\lambda>} a(x,\xi)dx. \tag{3.5}$$

As $a(x,\xi) \in \mathcal{S}(\mathbb{R}^n_x) \forall \xi \in \mathbb{R}^n/\{o\}$, it follows that $\tilde{a}(\lambda,\xi) \in \mathcal{S}(\mathbb{R}^n_\lambda) \forall \xi \in \mathbb{R}^n/\{o\}$ and the inversion formula

$$a(x,\xi) = (2\pi)^{n/2} \int_{\mathbb{R}^n} e^{i<x,\lambda>} \tilde{a}(\lambda,\xi)d\lambda, \ \forall (x,\xi) \in \mathbb{R}^n \times \mathbb{R}^n/\{o\} \tag{3.6}$$

– which is the definition of symbols in this Chapter – at (1.1), holds true.

As seen in Ch. VI -2, the function $\tilde{a}(\lambda,\xi)$ is continuous – hence measurable – in $\mathbb{R}^n \times \mathbb{R}^n/\{o\}$; the function $\lambda \in \mathbb{R}^n \to \tilde{a}(\lambda,\xi)$ is also continuous – hence measurable – $\forall \xi \in \mathbb{R}^n/\{o\}$; the function $k^1(\lambda) = \sup_{\xi \in \mathbb{R}^n/\{o\}} |\tilde{a}(\lambda,\xi)|$ is a function in $L^1(\mathbb{R}^n)$, and, even more,

$$(1+|\lambda|^2)^{|s|/2} k^1(\lambda) \in L^1(\mathbb{R}^n) \ \forall s \in \mathbb{R}. \tag{3.7}$$

This implies that $|\lambda| k^1(\lambda) \in L^1(\mathbb{R}^n)$ (take $s = 1$, and note that $|\lambda| k^1(\lambda) = (|\lambda|^2)^{1/2} k^1(\lambda) \leq (1+|\lambda|^2)^{1/2} k^1(\lambda) \in L^1(\mathbb{R}^n)$).

Next, as seen in Ch. VI (Prop. 1.3), the following estimates hold true:

$$(1+|\lambda|^2)^p |\tilde{a}(\lambda,\xi) - \tilde{a}(\lambda,\eta)| \leq C_p \frac{|\xi-\eta|}{|\xi|+|\eta|}, \ \forall \xi,\eta \in \mathbb{R}^n/\{o\}, \lambda \in \mathbb{R}^n, p = 0,1,2,\ldots \tag{3.8}$$

It we use also the inequality (2.24) – Ch. VI we get from (3.8), that

$$|\tilde{a}(\lambda,\xi) - \tilde{a}(\lambda,\eta) \leq C_p (1+|\lambda|^2)^{-p} (1+|\xi-\eta|^2)^{1/2} (1+\eta|^2)^{-1/2}, \ \forall \lambda \in \mathbb{R}^n, \xi,\eta \in \mathbb{R}^n/\{o\} \tag{3.9}$$

or, also changing η with ξ in (3.9):

$$|\tilde{a}(\lambda,\xi) - \tilde{a}(\lambda,\eta) \leq C_p (1+|\lambda|^2)^{-p} (1+|\xi-\eta|^2)^{1/2} (1+\xi|^2)^{-1/2}, \ \forall \lambda \in \mathbb{R}^n, \xi,\eta \in \mathbb{R}^n/\{o\} \tag{3.10}$$

Furthermore, let us define a function $\varphi(\cdot)$ by:

$\varphi(t) = \frac{1}{\sqrt{1+t^2}}, t \geq 0$; then φ is continuous on $[0,\infty), \varphi(t) > 0 \forall t \geq 0$, and $\varphi(t) \to 0$ as $t \to \infty$.

Next, note that also: $(1+x^2)^{1/2} \leq 1+x, \forall x \geq 0$, hence: $(1+|\xi-\eta|^2)^{1/2} \leq 1+|\xi-\eta|, \forall \xi,\eta \in \mathbb{R}^n$.

Hence, by now we have, using (3.10) and previous remarks, inequality:

$$|\tilde{a}(\lambda,\xi) - \tilde{a}(\lambda,\eta) \leq C_p(1+|\lambda|^2)^{-p}(1+|\xi-\eta|)\varphi(|\xi|), \forall \lambda \in \mathbb{R}^n, \xi,\eta \in \mathbb{R}^n/\{o\} \quad (3.11)$$

Accordingly one obtains:

$$\frac{|\tilde{a}(\lambda,\xi) - \tilde{a}(\lambda,\eta)|}{(1+|\xi-\eta|)\varphi(|\xi|)} \leq C_p(1+|\lambda|^2)^{-p} \; \forall p = 0,1,2,\ldots, \lambda \in \mathbb{R}^n, \xi,\eta \in \mathbb{R}^n/\{o\}. \quad (3.12)$$

Let be

$$k^2(\lambda) = \sup_{\xi,\eta \in \mathbb{R}^n/\{o\}} \frac{|\tilde{a}(\lambda,\xi) - \tilde{a}(\lambda,\eta)|}{(1+|\xi-\eta|)\varphi(|\xi|)}, \; \forall \lambda \in \mathbb{R}^n \quad (3.13)$$

We first show that $k^2(\lambda)$ is a measurable function on $\mathbb{R}^n$:

Take $(\xi_j,\eta_j)_{j\in\mathbb{N}}$ the sequence of points with rational coordinates in $\mathbb{R}^n/\{o\} \times \mathbb{R}^n/\{o\}$

Then:

$$k^2(\lambda) = \sup_{j\in\mathbb{N}} \frac{|\tilde{a}(\lambda,\xi_j) - \tilde{a}(\lambda,\eta_j)|}{(1+|\xi_j-\eta_j|)\varphi(|\xi_j|)} = \sup_{j\in\mathbb{N}} K_j^2(\lambda) \quad (3.14)$$

In fact it is obvious that : $\sup K_j^2(\lambda) \leq k^2(\lambda, \forall \lambda \in \mathbb{R}^n$. Assume that $\exists \bar{\lambda} \in \mathbb{R}^n$, such that

$$\sup_{j\in\mathbb{N}} K_j^2(\bar{\lambda}) < k^2(\bar{\lambda}) \text{ (strict inequality)} \quad (3.15)$$

Then, $\exists (\xi,\eta) \in \mathbb{R}^n/\{o\} \times \mathbb{R}^n/\{o\}$, such that the lower bound

$$\frac{|\tilde{a}(\bar{\lambda},\xi) - \tilde{a}(\bar{\lambda},\eta)|}{(1+|\xi-\eta|)\varphi(|\xi|)} > \sup_{j\in\mathbb{N}} K_j^2(\bar{\lambda}) \quad (3.16)$$

holds true.

Take now a subsequence $(\xi_{j_p},\eta_{j_p})_{p=1}^{\infty}$ which is convergent to (ξ,η). From continuity we obtain (as $p \to \infty$):

$$\frac{|\tilde{a}(\bar{\lambda}, \xi_{j_p}) - \tilde{a}(\bar{\lambda}, \eta_{j_p})|}{(1 + |\xi_{j_p} - \eta_{j_p}|)\varphi(|\xi_{j_p}|)} = K^2_{j_p}(\bar{\lambda}) \text{ is convergent to } \frac{|\tilde{a}(\bar{\lambda}, \xi) - \tilde{a}(\bar{\lambda}, \eta)|}{(1 + |\xi - \eta|)\varphi(|\xi|)} >$$

$\sup_{j \in \mathbb{N}} K^2_j(\bar{\lambda})$; however, $K^2_{j_p}(\bar{\lambda}) \le \sup_{j \in \mathbb{N}} K^2_j(\bar{\lambda})$, hence

$$\lim_{p \to \infty} K^2_{j_p}(\bar{\lambda}) \le \sup_{j \in \mathbb{N}} K^2_j(\bar{\lambda}) \text{ , hence: } \frac{|\tilde{a}(\bar{\lambda}, \xi) - \tilde{a}(\bar{\lambda}, \eta)|}{(1 + |\xi - \eta|)\varphi(|\xi|)} \le \sup_{j \in \mathbb{N}} K^2_j(\bar{\lambda})$$

which contradicts (3.16). This gives equality (3.14).

Therefore, from continuity of functions: $\lambda \in \mathbb{R}^n \to K^2_j(\lambda)$, we get the measurable function $k^2(\lambda), \mathbb{R}^n \to \mathbb{C}$, such that (because of (3.12)):

$$k^2(\lambda) \le C_p(1 + |\lambda|^2)^{-p} \forall p = 0, 1, 2, \ldots, \ \forall \lambda \in \mathbb{R}^n \tag{3.17}$$

This shows that: $(1 + |\lambda|)k^2(\lambda) \in L^1(\mathbb{R}^n)$.

Finally, we take: $k(\lambda) = \max\{k^1(\lambda), k^2(\lambda)\}, \forall \lambda \in \mathbb{R}^n$; it is again measurable, and such that $(1+|\lambda|)k(\lambda) \in L^1(\mathbb{R}^n)$; thus, both conditions i)–ii) are satisfied with this $k(\lambda)$ and $\varphi(t) = \frac{1}{\sqrt{1+t^2}}, \forall t \ge 0$.

It remains therefore to check condition iii).

First we have (from (1.5) – Ch. VI), the sequence of estimates:

$$(1 + |x|^2)^p |a(x, \xi)| \le C_p \ \forall x \in \mathbb{R}^n, \xi \in \mathbb{R}^n / \{o\} \quad p = 0, 1, 2 \ldots \tag{3.18}$$

hence $|a(x, \xi)| \le C_p(1 + |x|^2)^{-p}, \forall p \in \mathbb{N}$ and $|x| \ |a(x, \xi)| \le C_p \frac{|x|}{(1+|x|^2)^p}, \forall p \in \mathbb{N}$; if p is large enough we get

$$\int_{\mathbb{R}^n} |x| \ |a(x, \xi)| dx \le \Gamma, \ \forall \xi \in \mathbb{R}^n / \{o\} \tag{3.19}$$

Finally, let us examine condition (1.6) in assumption iii):

$$\lim_{|h| \to 0} \int_{|\xi| \le \Lambda} \left(\int_{\mathbb{R}^n} |a(x, \xi + h) - a(x, \xi)| dx \right) d\xi = 0 \ \forall \Lambda > 0.$$

We use Corollary 1.2 to Prop. 1.3 (Ch. VI) to write (using (1.33)):

$$|a(x, \xi + h) - a(x, \xi)| \le C_q \frac{|h|}{|\xi + h| + |\xi|} (1 + |x|^2)^{-q}, \ \forall \xi \in \mathbb{R}^n / \{o\},$$
$$\xi \ne -h, h \in \mathbb{R}^n, x \in \mathbb{R}^n \ q = 0, 1, 2, \ldots \tag{3.20}$$

This entails (taking large q), that

$$\int_{\mathbb{R}^n} |a(x,\xi+h)-a(x,\xi)|dx \le \frac{|h|}{|\xi|},\ \forall \xi \in \mathbb{R}^n/\{o\},\ h \in \mathbb{R}^n, \xi \ne -h \tag{3.21}$$

On the other hand, from: $|a(x,\xi)| \le C_q(1+|x|^2)^{-q}$, $\forall x \in \mathbb{R}^n, \xi \in \mathbb{R}^n/\{o\}$, $q = 0,1,2\ldots$, we derive the estimate:

$$\int_{\mathbb{R}^n} |a(x,\xi+h)-a(x,\xi)|dx \le \int_{\mathbb{R}^n} |a(x,\xi)|dx + \int_{\mathbb{R}^n} |a(x,\xi+h)|dx$$
$$\le C,\ \forall \xi \in \mathbb{R}^n/\{o\},\ \forall h \in \mathbb{R}^n, \xi \ne -h \tag{3.22}$$

We now write, $\forall \Lambda > 0$, the obvious equality:

$$\int_{|\xi|\le\Lambda} \left(\int_{\mathbb{R}^n} |a(x,\xi+h)-a(x,\xi)|dx\right) = \int_{|\xi|\le\sqrt{|h|}} \left(\int_{\mathbb{R}^n} |a(x,\xi+h)-a(x,\xi)|dx\right) d\xi$$
$$+ \int_{\sqrt{|h|}\le|\xi|\le\Lambda} \left(\int_{\mathbb{R}^n} |a(x,\xi+h)-a(x,\xi)|dx\right) d\xi = I_1 + I_2 \tag{3.23}$$

(we assume $|h| < \Lambda$ which can be done).

We next estimate I_1 using (3.22) so that

$$I_1 \le C \text{ meas } B(0,\sqrt{|h|}) \tag{3.24}$$

We also estimate I_2 using (3.21) and obtain

$$I_2 \le \int_{\sqrt{|h|}\le|\xi|\le\Lambda} \frac{|h|}{|\xi|} d\xi \le \frac{|h|}{\sqrt{|h|}} \text{ meas } B(0,\Lambda) = \sqrt{|h|} \text{ meas } B(0,\Lambda) \tag{3.25}$$

where $B(0,r)$ means here the ball of centre 0 and radius r in $\mathbb{R}^n$.

Obviously then, $\lim_{|h|\to 0} (C \text{ meas } B(0,\sqrt{|h|}) + \sqrt{|h|} \text{ meas } B(0,\Lambda)) = 0$ and this gives the desired result: ((1.6) in iii) holds true).

■

Chapter VIII
Gohberg's Lemma and applications

Introduction

We shall first refer ourselves to Ch. VI, concerning the pseudo-differential operators $\mathcal{A}(x,D)$ and $A(x,D)$ which are associated with the – so called – Kohn–Nirenberg homogeneous of degree 0 and $C^\infty(\mathbb{R}^n \times \mathbb{R}^n/\{o\})$ symbols $a(x,\xi)$. As seen in Ch. VI these are linear continuous operators, $H^s(\mathbb{R}^n) \to H^s(\mathbb{R}^n)$, $\forall s \in \mathbb{R}$. Estimates

$$\| \mathcal{A}(x,D)u \|_{H^s} \leq C_s \| u \|_{H^s} , \| A(x,D)u \|_{H^s} \leq C_s \| u \|_{H^s} \tag{0.1}$$
$$\forall u \in H^s(\mathbb{R}^n) \text{ (and } \forall s \in \mathbb{R})$$

are therefore satisfied.

Using concepts explained in Ch.I we can say that operators $\mathcal{A}(x,D)$ and $A(x,D)$ have order 0.

($\lambda = 0$ belongs to their order set).

In the present Chapter we present a method (sometimes called "Gohberg's Lemma"), which permits to establish that the *true order* (as defined in Ch.I, 1.8) of both operators above is also equal to 0 – unless $a(x,\xi) \equiv 0$).

Precisely, we give here a more general form of this "Lemma" which in fact applies to operators $\mathcal{A}(x,D)$ which are associated to more general classes of symbols $a(x,\xi)$. We next prove that the true order of both operators $\mathcal{A}(x,D)$ and $A(x,D)$ equals 0, unless

$$\limsup_{|\xi|\to\infty} |a(x,\xi)| = 0 \ \forall x \in \mathbb{R}^n \tag{0.2}$$

Some related results, of independent interest will also be given.

1.

Let us consider measurable functions $g(x,\xi), \mathbb{R}^n \times \mathbb{R}^n/\{o\} \to \mathbb{C}$, admitting a representation formula

$$g(x,\xi) = (2\pi)^{-n/2} \int_{\mathbb{R}^n} e^{i<x,\lambda>} \gamma(\lambda,\xi) d\lambda + \tilde{g}(\xi), \ \forall (x,\xi) \in \mathbb{R}^n \times \mathbb{R}^n/\{o\} \tag{1.1}$$

where the function $\tilde{g}(\xi), \mathbb{R}^n/\{o\} \to \mathbb{C}$ is bounded and measurable, while the function

$\lambda \to \gamma(\lambda,\xi),\ \lambda \in \mathbb{R}^n$, is measurable $\forall \xi \in \mathbb{R}^n/\{o\}$ and

the function $(\lambda,\xi) \to \gamma(\lambda,\xi)$ is also measurable, $\mathbb{R}^n \times \mathbb{R}^n/\{o\} \to \mathbb{C}$. We shall furthermore assume that

$$|\gamma(\lambda,\xi)| \le k(\lambda),\ \forall \lambda \in \mathbb{R}^n, \xi \in \mathbb{R}^n/\{o\} \tag{1.2}$$

where

$$k(\lambda) \in L^1(\mathbb{R}^n) \text{ and } (1+|\lambda|^2)^{|s|/2} k(\lambda) \in L^1(\mathbb{R}^n) \forall s \in \mathbb{R} \tag{1.3}$$

and also that a continuous function $\varphi(t), [0,\infty) \to]0,\infty)$ exists, such that

$$\lim_{t\to\infty} \sqrt{t}\varphi(t) = 0 \tag{1.4}$$

and the estimate

$$|\gamma(\lambda,\xi) - \gamma(\lambda,\eta)| \le (1+|\xi-\eta|)\varphi(|\xi|)k(\lambda) \tag{1.5}$$
$$\forall \xi,\eta \in \mathbb{R}^n/\{o\}, \forall \lambda \in \mathbb{R}^n$$

holds true, as well as the estimate

$$|\tilde{g}(\xi) - \tilde{g}(\eta)| \le (1+|\xi-\eta|)\varphi(|\xi|), \forall \xi,\eta \in \mathbb{R}^n/\{o\} \tag{1.6}$$

As seen in Ch.IV (section 2), the operator $\mathcal{G}(x,D)$ associated to $g(x,\xi)$ via the representation formula

$$\mathcal{G}(x,D)U = \mathcal{F}^{-1}[(2\pi)^{-n/2}\int_{\mathbb{R}^n} \gamma(\xi-\eta,\eta)\hat{U}(\eta)d\eta)] + \mathcal{F}^{-1}[\tilde{g}(\xi)\hat{U}(\xi)] \tag{1.7}$$
$$\forall U \in H^s(\mathbb{R}^n)$$

is a well-defined linear continuous mapping, $H^s(\mathbb{R}^n) \to H^s(\mathbb{R}^n)$ and, according to Ch.V – Prop. 1.2, it admits the alternative representation formula

$$(\mathcal{G}(x,D)u)(x) = (2\pi)^{-n/2}\int_{\mathbb{R}^n} e^{i<x,\eta>} g(x,\eta)\hat{u}(\eta)d\eta,\ \forall u \in \mathcal{S}(\mathbb{R}^n),\ \forall x \in \mathbb{R}^n \tag{1.8}$$

Remarks a) the integral in (1.8) is absolutely convergent: we see the upper bound for the symbol g (from (1.1)):

$$|g(x,\eta)| \leq (2\pi)^{-n/2} \int_{\mathbb{R}^n} |\gamma(\lambda,\eta)| d\lambda + |\tilde{g}(\eta)| \leq (2\pi)^{-n/2} \int_{\mathbb{R}^n} k(\lambda) d\lambda + \sup |\tilde{g}(\eta)|,$$

$$(x,\eta) \in \mathbb{R}^n \times \mathbb{R}^n/\{o\} \tag{1.9}$$

and use the fact that $\hat{u}(\cdot) \in \mathcal{S}(\mathbb{R}^n)$.

b) the function

$x \in \mathbb{R}^n \to (\mathcal{G}(x,D)u)(x)$ is bounded over $\mathbb{R}^n$, as follows from

$$|(\mathcal{G}(x,D)u)(x)| \leq (2\pi)^{-n/2} \sup |g(x,\eta)| \int_{\mathbb{R}^n} |\hat{u}(\eta)| d\eta$$

c) The function in b) is also continuous on $\mathbb{R}^n$; in fact, the function $x \to g(x,\eta), x \in \mathbb{R}^n$, is continuous $\forall \eta \in \mathbb{R}^n/\{o\}$ – it follows from (1.1); hence, the function: $x \to e^{i<x,\eta>} g(x,\eta)\hat{u}(\eta)$ is also continuous, $\forall x \in \mathbb{R}^n, \forall \eta \in \mathbb{R}^n/\{o\}$. Applying the dominated convergence theorem in (1.8) we find that $x \to (\mathcal{G}(x,D)u)(x)$ is continuous, $\forall x \in \mathbb{R}^n$.

We can conclude from above observations that, if $u(\cdot) \in \mathcal{S}(\mathbb{R}^n)$ then $(\mathcal{G}(x,D)u)(\cdot) \in C_b(\mathbb{R}^n)$ (the bounded continous functions, $\mathbb{R}^n \to \mathbb{C}$).

2.

In the subsequent discussion we shall make an essential usage of the linear operators $P_{x_0,\xi_0}, Q_{x_0,\xi_0}$ (depending on the point (x_0,ξ_0) in $\mathbb{R}^n \times \mathbb{R}^n/\{o\}$), which are – formally – defined on functions $u(\cdot), \mathbb{R}^n \to \mathbb{C}$ by means of the relations:

$$(P_{x_0,\xi_0}u)(x) = |\xi_0|^{n/4} u((x - x_0)|\xi_0|^{1/2}) e^{i<x,\xi_0>} \tag{2.1}$$

$$(Q_{x_0,\xi_0}u)(x) = |\xi_0|^{-n/4} u((x_0 + |\xi_0|^{-1/2} x) e^{-i<x_0+|\xi_0|^{-1/2}x,\xi_0>} \tag{2.2}$$

where $x_0 \in \mathbb{R}^n, \xi_0 \in \mathbb{R}^n/\{o\}$ are fixed.

We now state

Proposition 2.1 *The operators P_{x_0,ξ_0} and Q_{x_0,ξ_0} are linear mappings*

$$\mathcal{S}(\mathbb{R}^n) \to \mathcal{S}(\mathbb{R}^n); C_b(\mathbb{R}^n) \to C_b(\mathbb{R}^n); L^2(\mathbb{R}^n) \to L^2(\mathbb{R}^n)$$

$$\forall (x_0,\xi_0) \in \mathbb{R}^n \times \mathbb{R}^n/\{o\}.$$

Proof First of all, it is obvious that all continuous functions on $\mathbb{R}^n$ are mapped by the above operators P and Q – again – in continuous functions. We also have estimates:

$$|(P_{x_0,\xi_0}u)(x)| \leq |\xi_0|^{n/4} \sup_{\mathbb{R}^n} |u(y)|, |(Q_{x_0,\xi_0}v)(x)| \leq |\xi_0|^{-n/4} \sup_{\mathbb{R}^n} |v(y)| \tag{2.3}$$

Thus, in fact, $C_b(\mathbb{R}^n)$ goes into itself, through the action of operators P and Q.

Furthermore, we readily see that if u, v are measurable on $\mathbb{R}^n$, then Pu, Qv are also measurable on $\mathbb{R}^n$; applying the change of variable theorem we obtain the equality:

$$\begin{aligned}\int_{\mathbb{R}^n} |(P_{x_0,\xi_0}u)(x)|^2 dx &= \int_{\mathbb{R}^n} |\xi_0|^{n/2} |u((x - x_0)|\xi_0|^{1/2})|^2 dx = (\text{if } (x - x_0)|\xi_0|^{1/2} = y) \\ &= \int_{\mathbb{R}^n} |u(y)|^2 dy \end{aligned} \tag{2.4}$$

and also

$$\begin{aligned}\int_{\mathbb{R}^n} |(Q_{x_0,\xi_0}v)(x)|^2 dx &= \int_{\mathbb{R}^n} |\xi_0|^{-n/2} |v(x_0 + |\xi_0|^{-1/2}x)|^2 dx = \\ (\text{if } x_0 + |\xi_0|^{-1/2}x &= y, dy = |\xi_0|^{-n/2} dx) \\ &= \int_{\mathbb{R}^n} |v(y)|^2 dy \end{aligned} \tag{2.5}$$

thus, if $u, v \in L^2(\mathbb{R}^n)$ then Pu, Qv also belong to $L^2(\mathbb{R}^n)$; they are *isometric* linear operators in this space.

Finally, let $u \in \mathcal{S}(\mathbb{R}^n)$; then the function $x \to u(x - x_0)$ also belongs to $\mathcal{S}(\mathbb{R}^n)$, as well as the functions $x \to u((x - x_0)|\xi_0|^{1/2})$ and then $x \to (P_{x_0,\xi_0}u)(x)$ (see [9] – p.49, 61,63).

For similar reasons, if $v \in \mathcal{S}(\mathbb{R}^n)$ then $Q_{x_0,\xi_0}v \in \mathcal{S}(\mathbb{R}^n)$ too.

■

Let us note also

Proposition 2.2 *The operators P, Q are inverse to each other, in any of the above spaces.*

(we omit the index (x_0, ξ_0)).

Proof We must therefore establish that $PQ = QP =$ Identity. We have in fact:

$(Qv)(x) = |\xi_0|^{-n/4}v(x_0 + |\xi_0|^{-1/2}x)e^{-i<x_0+|\xi_0|^{-1/2}x,\xi_0>}$. Therefore we get:

$$(Qv)((x - x_0)|\xi_0|^{1/2}) = |\xi_0|^{-n/4}v(x_0 + |\xi_0|^{-1/2}((x - x_0)|\xi_0|^{1/2}))$$
$$e^{-i<x_0+|\xi_0|^{-1/2}((x-x_0)|\xi_0|^{1/2}),\xi_0>} = |\xi_0|^{-n/4}v(x)e^{-i<x,\xi_0>}$$

and accordingly we obtain:

$$(PQv)(x) = |\xi_0|^{n/4}e^{i<x,\xi_0>}|\xi_0|^{-n/4}v(x)e^{-i<x,\xi_0>} = v(x). \tag{2.6}$$

Also we have
$(Pu)(x) = |\xi_0|^{n/4}u((x - x_0)|\xi_0|^{1/2})e^{i<x,\xi_0>}$; hence we obtain
$P(u)(x_0 + |\xi_0|^{-1/2}x) = |\xi_0|^{n/4}e^{i<x_0+|\xi_0|^{-1/2}x,\xi_0>}u(|\xi_0|^{1/2}|\xi_0|^{-1/2}x) = |\xi_0|^{n/4}e^{i<x_0,\xi_0>}e^{i<|\xi_0|^{-1/2}x,\xi_0>}u(x)$, and accordingly, it follows that

$$(QPu)(x) = |\xi_0|^{-n/4}e^{-i<x_0+|\xi_0|^{-1/2}x,\xi_0>}|\xi_0|^{n/4}e^{i<x_0,\xi_0>}e^{i<|\xi_0|^{-1/2}x,\xi_0>}u(x) = u(x). \tag{2.7}$$

(in the above computations v and u may belong to $\mathcal{S}(\mathbb{R}^n), C_b(\mathbb{R}^n)$ or $L^2(\mathbb{R}^n)$).

∎

Next, we shall give a formula for the action of the product operator $Q_{x_0,\xi_0}\mathcal{G}P_{x_0,\xi_0}$ on $u \in \mathcal{S}(\mathbb{R}^n)$.

Proposition 2.3 *If $u \in \mathcal{S}(\mathbb{R}^n)$ and $(x_0, \xi_0) \in \mathbb{R}^n \times \mathbb{R}^n/\{o\}$, it results that*

$$(Q_{x_0,\xi_0}\mathcal{G}P_{x_0,\xi_0})u(x) = (2\pi)^{-n/2}\int_{\mathbb{R}^n} e^{i<x,\xi>}g(x_0+|\xi_0|^{-1/2}x, \xi_0+\sqrt{|\xi_0|}\xi)\hat{u}(\xi)d\xi\forall x \in \mathbb{R}^n \tag{2.8}$$

Proof First we compute $\mathcal{G}P_{x_0,\xi_0}u$, for $u \in \mathcal{S}$, using (1.8) and (2.1). We thus have

$$(\mathcal{G}(x, D)P_{x_0,\xi_0}u)(x) = (2\pi)^{-n/2}\int_{\mathbb{R}^n} e^{i<x,\eta>}g(x, \eta)(P_{x_0,\xi_0}u)\hat{}(\eta)d\eta, \forall u \in \mathcal{S}(\mathbb{R}^n). \tag{2.9}$$

Actually, the Fourier transform of Pu is given by:

$$(P_{x_0,\xi_0}u)\hat{}(\eta) = (2\pi)^{-n/2}\int_{\mathbb{R}^n} e^{-i<x,\eta>}|\xi_0|^{n/4}u(|\xi_0|^{1/2}(x - x_0))e^{i<x,\xi_0>}dx =$$

(after the substitution $(x - x_0)|\xi_0|^{1/2} = y$) :

$$(2\pi)^{-n/2}\int_{\mathbb{R}^n} e^{-i<x_0+|\xi_0|^{-1/2}y,\eta>}|\xi_0|^{n/4}u(y)e^{i<x_0+|\xi_0|^{-1/2}y,\xi_0>}|\xi_0|^{-n/2}dy =$$

$$(2\pi)^{-n/2}|\xi_0|^{-n/4}\int_{\mathbb{R}^n} e^{-i<x_0+|\xi_0|^{-1/2}y,\eta-\xi_0>}u(y)dy = (2\pi)^{-n/2}|\xi_0|^{-n/4}e^{-i<x_0,\eta-\xi_0>}$$

$$\int_{\mathbb{R}^n} e^{-i<y,|\xi_0|^{-1/2}(\eta-\xi_0)>}$$

$$u(y)dy = |\xi_0|^{-n/4} e^{-i<x_0,\eta-\xi_0>} \hat{u}(\frac{\eta-\xi_0}{\sqrt{|\xi_0|}}) \tag{2.10}$$

Accordingly, the formula for $\mathcal{G}(x,D)P_{x_0,\xi_0}u(x)$ becomes:

$$(\mathcal{G}Pu)(x) = (2\pi)^{-n/2} \int_{\mathbb{R}^n} e^{i<x,\eta>} g(x,\eta)|\xi_0|^{-n/4} e^{-i<x_0,\eta-\xi_0>} \hat{u}(\frac{\eta-\xi_0}{\sqrt{|\xi_0|}}) d\eta \tag{2.11}$$

where, by previous remarks, $\mathcal{G}Pu \in C_b(\mathbb{R}^n)$.

Finally, we compute $Q\mathcal{G}Pu$ using (2.2); we see that $Q\mathcal{G}Pu \in C_b(\mathbb{R}^n)$ too. We have

$$(Q\mathcal{G}Pu)(x) = |\xi_0|^{-n/4} e^{-i<x_0+|\xi_0|^{-1/2}x,\xi_0>} (\mathcal{G}Pu)(x_0+|\xi_0|^{-1/2}x) =$$

$$|\xi_0|^{-n/4} e^{-i<x_0+\xi_0|^{-1/2}x,\xi_0>} (2\pi)^{-n/2} \int_{\mathbb{R}^n} e^{i<x_0+|\xi_0|^{-1/2}x,\eta>} g(x_0+|\xi_0|^{-1/2}x,\eta) \cdot |\xi_0|^{-n/4} \cdot$$

$$e^{-i<x_0,\eta-\xi_0>} \hat{u}(\frac{\eta-\xi_0}{\sqrt{|\xi_0|}}) d\eta =$$

$$(2\pi)^{-n/2} \int_{\mathbb{R}^n} e^{i<|\xi_0|^{-1/2}x,\eta-\xi_0>} g(x_0+|\xi_0|^{-1/2}x,\eta)|\xi_0|^{-n/2} \hat{u}(\frac{\eta-\xi_0}{\sqrt{|\xi_0|}}) d\eta$$

$=$ (after substitution $(\eta-\xi_0)|\xi_0|^{-1/2} = \xi, d\eta = |\xi_0|^{n/2} d\xi$:

$$= (2\pi)^{-n/2} \int_{\mathbb{R}^n} e^{i<x,\xi>} g(x_0+|\xi_0|^{-1/2}x, \xi|\xi_0|^{1/2}+\xi_0) \hat{u}(\xi) d\xi \tag{2.12}$$

$$\forall x \in \mathbb{R}^n.$$

This establishes (2.8), as requested.

3.

In this section we shall demonstrate a certain result about the above studied product operator $Q\mathcal{G}P$ which in various occasions appears quite useful (see for instance [14] – p.295-296).

Theorem 3.1 *Consider symbols $g(x,\xi)$ as in (1.1), satisfying (1.2)-(1.3) with $s=1$ - (1.4), (1.5), (1.6)*

Assume that, for some $x_0 \in \mathbb{R}^n$ and a sequence (ξ_p) in $\mathbb{R}^n/\{o\}$ with $|\xi_p| \to \infty$, we have

$$\lim_{p\to\infty} g(x_0,\xi_p) = C_0 \in \mathbb{C}. \tag{3.1}$$

It follows that:

$$lim_{p\to\infty}(Q_{x_0,\xi_p}\mathcal{G}(x,D)P_{x_0,\xi_p}u)(x) = C_0 u(x) \tag{3.2}$$

holds, $\forall u \in \mathcal{S}(\mathbb{R}^n)$*, uniformly on bounded sets of* $\mathbb{R}^n$*.*

Proof We start with the obvious equality

$$C_0 u(x) = (2\pi)^{-n/2} \int_{\mathbb{R}^n} C_0 e^{i<x,\eta>}\hat{u}(\eta)d\eta, \ \forall u \in \mathcal{S} \tag{3.3}$$

Accordingly we obtain (using (2.8) and (3.3)):

$$\begin{aligned} &(Q_{x_0,\xi_p}\mathcal{G}(x,D)P_{x_0,\xi_p}u)(x) - C_0 u(x) = \\ &(2\pi)^{-n/2} \int_{\mathbb{R}^n} e^{i<x,\eta>}[g(x_0 + |\xi_p|^{-1/2}x, \xi_p + |\xi_p|^{1/2}\eta) - C_0]\hat{u}(\eta)d\eta \end{aligned} \tag{3.4}$$

We see also that (for $\eta \neq -|\xi_p|^{-1/2}\xi_p$ where $g(x_0 + |\xi_p|^{-1/2}x, \xi_p + |\xi_p|^{1/2}\eta)$ is undefined) $g(x_0 + |\xi_p|^{-1/2}x, \xi_p + |\xi_p|^{1/2}\eta) - C_0 = g(x_0 + |\xi_p|^{-1/2}x, \xi_p + |\xi_p|^{1/2}\eta) - g(x_0, \xi_p) + g(x_0, \xi_p) - C_0$ hence the relation (3.4) entails:

$$\begin{aligned} &(Q_{x_0,\xi_p}\mathcal{G}(x,D)P_{x_0,\xi_p}u)(x) - C_0 u(x) = \\ &(2\pi)^{-n/2} \int_{\mathbb{R}^n} e^{i<x,\eta>}[g(x_0 + |\xi_p|^{-1/2}x, \xi_p + |\xi_p|^{1/2}\eta) - g(x_0, \xi_p)]\hat{u}(\eta)d\eta + \\ &(2\pi)^{-n/2} \int_{\mathbb{R}^n} e^{i<x,\eta>}[g(x_0, \xi_p) - C_0]\hat{u}(\eta)d\eta = I_1 + I_2 \end{aligned} \tag{3.5}$$

which are absolutely convergent integrals (as $\hat{u} \in \mathcal{S}$).

The second integral in the right-hand side of (3.5) is obviously estimated in absolute value by

$$(2\pi)^{-n/2}|g(x_0, \xi_p) - C_0| \int_{\mathbb{R}^n} |\hat{u}(\eta)|d\eta$$

which $\to 0$ as $p \to \infty$ (uniformly for $x \in \mathbb{R}^n$).

Next, in order to evaluate (in absolute value) the first integral in the right-hand side of (3.5), we shall start with the obvious decomposition:
(for $\eta \neq -|\xi_p|^{-1/2}\xi_p$)

$$\begin{aligned} &g(x_0 + |\xi_p|^{-1/2}x, \xi_p + |\xi_p|^{1/2}\eta) - g(x_0, \xi_p) = g(x_0 + |\xi_p|^{-1/2}x, \xi_p + |\xi_p|^{1/2}\eta) - \\ &g(x_0, \xi_p + |\xi_p|^{1/2}\eta) + g(x_0, \xi_p + |\xi_p|^{1/2}\eta) - g(x_0, \xi_p) \end{aligned} \tag{3.6}$$

We shall need in the following a few more results on the symbols, expressed as

Lemma 3.1 *For the class of symbols in Th. 3.1 we have also:*

a) $$\frac{\partial g}{\partial x_j} \text{ exists, } \forall j = 1, 2, \dots n \quad \text{and belongs to} \quad C_b(\mathbb{R}^n) \tag{3.7}$$

b) $$|g(x,\xi) - g(y,\xi) \leq C|x-y|, \ \forall x, y \in \mathbb{R}^n, \xi \in \mathbb{R}^n/\{o\} \tag{3.8}$$

holds, with an absolute constant C.

In fact, to establish a) we use formula 1.1); we have the estimates

$$|e^{i<x,\lambda>}\gamma(\lambda,\xi)| \leq |\gamma(\lambda,\xi)| \leq k(\lambda) \in L^1(\mathbb{R}^n), \ \forall \xi \in \mathbb{R}^n/\{o\} \tag{3.9}$$

$$|i\lambda_j e^{i<x,\lambda>}\gamma(\lambda,\xi)| \leq |\lambda j|k(\lambda) \leq |\lambda|k(\lambda) \in L^1(\mathbb{R}^n), \ \forall \xi \in \mathbb{R}^n/\{o\} \tag{3.10}$$

It results

$\frac{\partial g}{\partial x_j}$ exists and $= (2\pi)^{-n/2} \int_{\mathbb{R}^n} e^{i<x,\lambda>}(i\lambda_j)\gamma(\lambda,\xi)d\lambda$, a continuous function of $x \in \mathbb{R}^n$;

furthermore

$$\left|\frac{\partial g}{\partial x_j}\right| \leq (2\pi)^{-n/2} \int_{\mathbb{R}^n} |\lambda|k(\lambda)d\lambda, \text{ hence } \frac{\partial g}{\partial x_j} \in C_b(\mathbb{R}^n)$$

In order to establish the (Lipschitz) estimate b) we see that (by the mean value theorem in $\mathbb{R}^n_x$):

$$g(x,\xi) - g(y,\xi) = \sum_{j=1}^{n}(x_j - y_j)\frac{\partial g}{\partial x_j}(z) \tag{3.11}$$

where z (which could depend on ξ), lays in the interval $[x, y]$. Accordingly:

$$|g(x,\xi) - g(y,\xi)| \leq |x-y|C_1 \tag{3.12}$$

where C_1 depends on the supremum of $\left|\frac{\partial g}{\partial x_j}\right|$ in $\mathbb{R}^n$, $\forall j = 1, 2, \dots n$.

$$\left(C_1 \geq [\sum_1^n (\frac{\partial g}{\partial x_j})^2]^{1/2} \text{ in } \mathbb{R}^n\right)$$

We continue now (and conclude) the *Proof of Th. 3.1.*

We consider the first integral (derived from (3.5) – and (3.6))

$$(2\pi)^{-n/2}\int_{\mathbb{R}^n} e^{i<x,\eta>}[g(x_0+|\xi_p|^{-1/2}x,\xi_p+|\xi_p|^{1/2}\eta)-g(x_0,\xi_p+|\xi_p|^{1/2}\eta)]\hat{u}(\eta)d\eta = I_3.$$

We estimate its absolute value using b) in Lemma: we get:

$$|I_3| \leq (2\pi)^{-n/2}|\xi_p|^{-1/2}|x|\cdot C\int_{\mathbb{R}^n}|\hat{u}(\eta)|d\eta \tag{3.13}$$

We see that this expression $\to 0$ as $p \to \infty$, uniformly for x in bounded sets of $\mathbb{R}^n$.

Finally, we look at the integral

$$I_4 = (2\pi)^{-n/2}\int_{\mathbb{R}^n} e^{i<x,\eta>}[g(x_0,\xi_p+|\xi_p|^{1/2}\eta)-g(x_0,\xi_p)]\hat{u}(\eta)d\eta \tag{3.14}$$

Let us use assumptions (1.3)–(1.5)–(1.6). We get, from (1.1), the following:

$$\begin{aligned}&|g(x,\xi)-g(x,\eta)| \leq (2\pi)^{-n/2}\int_{\mathbb{R}^n}|\gamma(\lambda,\xi)-\gamma(\lambda,\eta)|d\lambda + |\tilde{g}(\xi)-\tilde{g}(\eta)| \leq \\ &(2\pi)^{-n/2}\int_{\mathbb{R}^n}(1+|\xi-\eta|)\varphi(|\xi|)k(\lambda)d\lambda + (1+|\xi-\eta|)\varphi(|\xi|) = \\ &C_2(1+|\xi-\eta|)\varphi(|\xi|)+(1+|\xi-\eta|)\varphi(|\xi|) = \Gamma(1+|\xi-\eta|)\varphi(|\xi|), \xi,\eta \in \mathbb{R}^n/\{o\} \\ &x \in \mathbb{R}^n.\end{aligned} \tag{3.15}$$

It follows obviously (applying in I_4 with $\xi \to \xi_p, \eta \to \xi_p + |\xi_p|^{1/2}\eta$):

$$\begin{aligned}&|I_4| \leq (2\pi)^{-n/2}\Gamma\int_{\mathbb{R}^n}(1+|\xi_p|^{1/2}|\eta|)\varphi(|\xi_p|)|\hat{u}(\eta)|d\eta = \\ &C_3\varphi(|\xi_p|)\int_{\mathbb{R}^n}|\hat{u}(\eta)|d\eta + C_4\varphi(|\xi_p|)\sqrt{|\xi_p|}\int_{\mathbb{R}^n}|\eta||\hat{u}(\eta)|d\eta\end{aligned} \tag{3.16}$$

As $\sqrt{t}\varphi(t) \to 0$ for $t \to \infty$, it results also that $\varphi(t) \to 0$ as $t \to \infty$. Thus $I_4 \to 0$ as $p \to \infty$, and the estimate for I_4 is independent of $x \in \mathbb{R}^n$.

4.

In this section we extend previous considerations to operators acting in Sobolef spaces $H^s(\mathbb{R}^n)$ as well as to estimates in the $\| \ \|_s$-norm, where $s \in \mathbb{R}$. (this is approaching us to the main result of this Chapter, where the true order of the operator $\mathcal{G}(x,D)$ is seen to be 0 unless the symbol $g(x,\xi)$ is "essentially" null).

An initial result appears as

Proposition 4.1 *Let $x_0 \in \mathbb{R}^n$ and the sequence (ξ_p) in $\mathbb{R}^n$, with $|\xi_p| \to +\infty$ Then, $\forall \mathcal{E} > 0$, we have $\lim_{p\to\infty} P_{x_0,\xi_p} u = 0$ in $H_{-\mathcal{E}}$-norm , $\forall u \in \mathcal{S}(\mathbb{R}^n)$*

(remember that, $\forall s \in \mathbb{R}, \| u \|_{H^s} = \left(\int_{\mathbb{R}^n} (1+|\xi|^2)^s |\hat{u}(\xi)|^2 d\xi\right)^{1/2}$, $\forall u \in \mathcal{S}(\mathbb{R}^n)$).

Proof Use formula (2.1) and then (2.10). We get, for $u \in \mathcal{S}(\mathbb{R}^n)$, the equality

$$(P_{x_0,\xi_p} u)^\wedge(\xi) = |\xi_p|^{-n/4} e^{-i<x_0,\xi-\xi_p>} \hat{u}\left(\frac{\xi-\xi_p}{\sqrt{|\xi_p|}}\right) \quad (4.1)$$

Consequently we obtain, for $\mathcal{E} > 0$:

$$\int_{\mathbb{R}^n} (1+|\xi|^2)^{-\mathcal{E}} |(P_{x_0,\xi_p} u)^\wedge(\xi)|^2 d\xi = \int_{\mathbb{R}^n} (1+|\xi|^2)^{-\mathcal{E}} |\xi_p|^{-n/2} |\hat{u}\left(\frac{\xi-\xi_p}{\sqrt{|\xi_p|}}\right)|^2 d\xi =$$

$$\text{(after substitution } \frac{\xi-\xi_p}{\sqrt{|\xi_p|}} = \eta) : \int_{\mathbb{R}^n} (1+|\sqrt{|\xi_p|}\eta + \xi_p|^2)^{-\mathcal{E}} |\hat{u}(\eta)|^2 d\eta \quad (4.2)$$

Note now the estimate

$$|\xi_p + \sqrt{|\xi_p|}\eta| \ge |\xi_p| - \sqrt{|\xi_p|}|\eta| = \sqrt{|\xi_p|}(\sqrt{|\xi_p|} - |\eta|) \quad (4.3)$$

It follows that, $\forall \eta \in \mathbb{R}^n$ –fixed– , $\lim_{p\to\infty} |\xi_p + \sqrt{|\xi_p|}\eta| = +\infty$, hence $\lim_{p\to\infty}(1 + |\sqrt{|\xi_p|}\eta + \xi_p|^2)^{-\mathcal{E}} = 0, \forall \eta \in \mathbb{R}^n (\mathcal{E} > 0)$. Applying the dominated convergence theorem in (4.2) we get the result.

Next we shall define the mapping $I_s (\forall s \in \mathbb{R})$, from $\mathcal{S}(\mathbb{R}^n)$ into itself. Note that, $\forall s \in \mathbb{R}$, the function $\xi \in \mathbb{R}^n \to (1+|\xi|^2)^{s/2}$ is in $C^\infty(\mathbb{R}^n)$ and is slowly (polynomially) increasing together with all its derivatives. It follows (see for instance [9] p.61, 62) that the multiplication operator by $(1+|\xi|^2)^{s/2}$ maps $\mathcal{S}(\mathbb{R}^n)$ into itself.

Take $u \in \mathcal{S}(\mathbb{R}^n)$; its Fourier transform $\hat{u}(\xi)$ belongs to $\mathcal{S}(\mathbb{R}^n)$; thus $\xi \to (1+|\xi|^2)^{s/2}\hat{u}(\xi) \in \mathcal{S}(\mathbb{R}^n)$. Define:

$$(I_s u)(x) = \mathcal{F}^{-1}[(1+|\xi|^2)^{s/2}\hat{u}(\xi)] = (2\pi)^{-n/2} \int_{\mathbb{R}^n} e^{i<x,\xi>} (1+|\xi|^2)^{s/2} \hat{u}(\xi) d\xi \quad (4.4)$$

($\mathcal{F}^{-1}$ is the inverse Fourier transform, here acting in $\mathcal{S}(\mathbb{R}^n)$). We see that

$(I_s u)^\wedge(\xi) = (1+|\xi|^2)^{s/2}\hat{u}(\xi), \forall \xi \in \mathbb{R}^n$.

(another notation: $I_s = (I - \Delta)^{s/2}$, where $\Delta = \sum_{i=1}^n \frac{\partial^2}{\partial x_i^2}$; if $s = 2m$, an even integer, it is the usual (partial) differential operator).

Definition 4.1 *For* $(x_0, \xi_0) \in \mathbb{R}^n \times \mathbb{R}^n/\{o\}$ *and*

$$\forall s \in \mathbb{R}, P^s_{x_0,\xi_0} = (I - \Delta)^{-s/2} P_{x_0,\xi_0} \tag{4.5}$$

as an operator $\mathcal{S}(\mathbb{R}^n) \to \mathcal{S}(\mathbb{R}^n)$.

Remark We obviously have:

$$\| P^s_{x_0,\xi_0} u \|_{H^s} = \left(\int_{\mathbb{R}^n} (1+|\xi|^2)^s |(P^s_{x_0,\xi_0} u)^\wedge(\xi)|^2 d\xi\right)^{1/2} =$$

$$\left(\int_{\mathbb{R}^n} (1+|\xi|^2)^s (1+|\xi|^2)^{-s} |(P_{x_0,\xi_0} u)^\wedge(\xi)|^2 d\xi\right)^{1/2} = \| P_{x_0,\xi_0} u \|_{L^2} = \| u \|_{L^2}, \forall u \in \mathcal{S} \tag{4.6}$$

Note also the following

Corollary (to Prop. 4.1) *Under the assumptions of Prop. 4.1, it results that*

$$\forall \mathcal{E} > 0, \lim_{p\to\infty} \| P^s_{x_0,\xi_p} u \|_{H^{s-\varepsilon}} = 0, \ \forall u \in \mathcal{S}(\mathbb{R}^n) \tag{4.7}$$

This follows from Prop. 4.1 and the immediate relations:

$$\| I_{-s} P_{x_0,\xi_p} u \|^2_{H^{s-\varepsilon}} = \int_{\mathbb{R}^n} (1+|\xi|^2)^{s-\mathcal{E}} |(I_{-s} P_{x_0,\xi_p} u)^\wedge(\xi)|^2 d\xi =$$

$$\int_{\mathbb{R}^n} (1+|\xi|^2)^{s-\mathcal{E}} |(1+|\xi|^2)^{-s/2} (P_{x_0,\xi_p} u)^\wedge(\xi)|^2 d\xi = \int_{\mathbb{R}^n} (1+|\xi|^2)^{-\mathcal{E}} |(P_{x_0,\xi_p} u)^\wedge(\xi)|^2 d\xi \tag{4.8}$$

Next we give the (related to Defin. 4.1).

Definition 4.2 *For* $(x_0, \xi_0) \in \mathbb{R}^n \times \mathbb{R}^n/\{o\}$ *and* $\forall s \in \mathbb{R}, Q^s_{x_0,\xi_0} = Q_{x_0,\xi_0}$ $(I - \Delta)^{s/2}$, *an operator of* $\mathcal{S}(\mathbb{R}^n)$ *into itself.*

Remark We have,

$$\forall u \in \mathcal{S}(\mathbb{R}^n), \| Q^s_{x_0,\xi_0} u \|_{L^2(\mathbb{R}^n)} = \| (I-\Delta)^{s/2} u \|_{L^2(\mathbb{R}^n)} = \| u \|_{H^s(\mathbb{R}^n)} \tag{4.9}$$

Note also the relation, again for $u \in \mathcal{S}(\mathbb{R}^n)$:

$$P^s_{x_0,\xi_0} Q^s_{x_0,\xi_0} u = (I-\Delta)^{-s/2} P_{x_0,\xi_0} Q_{x_0,\xi_0} (I-\Delta)^{s/2} u = \text{ (by Prop. 2.2)}$$
$$(I-\Delta)^{-s/2}(I-\Delta)^{s/2} u = u \text{ (as readily seen!)} \tag{4.10}$$

In the following lines we extend the operator $(I-\Delta)^{s/2}$ as a (linear) operator from $\mathcal{S}'(\mathbb{R}^n)$ into itself. It is again the (Friedrichs) pseudo-differential operator $\psi_s(D)$ associated to the function $\xi \to (1+|\xi|^2)^{s/2}, \psi_s(\xi) = (1+|\xi|^2)^{s/2}$. Precisely, we put

$$\psi_s(D)T = (\mathcal{F}^{-1}\mathcal{M}_{\psi_s(\cdot)}\mathcal{F})T, \ \forall T \in \mathcal{S}'(\mathbb{R}^n). \tag{4.11}$$

(note that the slowly increasing C^∞ function $\psi_s(\xi)$ acts as a multiplicator on $\mathcal{S}'(\mathbb{R}^n)$ too).

Again, we shall use the notation

$$I_s = \mathcal{F}^{-1}\mathcal{M}_{\psi(\cdot)}\mathcal{F} \ ; \text{ we have: } I_s \ ; \ \mathcal{S}'(\mathbb{R}^n) \to \mathcal{S}'(\mathbb{R}^n).$$

We now establish

Proposition 4.2 *For any $\sigma, s \in \mathbb{R}$, the operator I_s is a linear isometric mapping of H^σ onto $H^{\sigma-s}$.*

Proof a) Let $T \in H^\sigma$; hence $\hat{T} \in L^1_{loc}(\mathbb{R}^n)$ and $(1+|\xi|^2)^{\sigma/2}\hat{T}(\xi) \in L^2(\mathbb{R}^n)$.

Consider I_sT (in $\mathcal{S}'$). Its Fourier transform: $\mathcal{F}(I_sT)$ is, accordingly, $\mathcal{M}_{\psi_s(\cdot)}(\mathcal{F}T) = \psi_s(\xi)\hat{T}(\xi) \in L^1_{loc}(\mathbb{R}^n)$; furthermore

$$(1+|\xi|^2)^{(\sigma-s)/2}\psi_s(\xi)\hat{T}(\xi) = (1+|\xi|^2)^{(\sigma-s)/2}(1+|\xi|^2)^{s/2}\hat{T}(\xi) = (1+|\xi|^2)^{\sigma/2}\hat{T}(\xi) \in L^2(\mathbb{R}^n)$$

b) Let $T \in H^{\sigma-s}$; define $U = I_{-s}T$; by a), $U \in H^{\sigma-s-(-s)} = H^\sigma$.

Also $I_sU = I_sI_{-s}T = T$, as readily seen.

c) From above we also see that, if $T \in H^\sigma$, then $\| I_sT \|_{H^{\sigma-s}} = \| T \|_{H^\sigma}$.

Remark 1 If $\sigma = 0$ we get: I_{-s} is linear isometric, H^o onto H^s.

Remark 2 The operator $P^s_{x_0,\xi_0} = I_{-s}P_{x_0,\xi_0}$ (see (4.5)) is linear isometric, H^o into H^s (see Prop. 2.1 and above remark).

We now make appeal to section 2, Ch. IV and obtain, under the present assumptions (1.1)–(1.2)–(1.3), that the operator $\mathcal{G}(x,D)$ associated to the symbol $g(x,\xi)$, is a linear continuous mapping, $H^s(\mathbb{R}^n) \to H^s(\mathbb{R}^n)$.

Then, the product operator:

$$\mathcal{G}(x,D)P^s_{x_0,\xi_0}$$

is linear continuous, $L^2(\mathbb{R}^n) \to H^s(\mathbb{R}^n)$.

Next, consider the operator $Q^s_{x_0,\xi_0} = Q_{x_0,\xi_0} I_s$; it is a linear continuous mapping, $H^s(\mathbb{R}^n) \to H^o(\mathbb{R}^n)$ (Prop. 4.2 with $\sigma = s$ and Prop. 2.1 again). Accordingly, we see that the operator $\mathcal{G}_s(x, D)$ defined by

$$\mathcal{G}_s(x, D) = Q^s_{x_0,\xi_0} \mathcal{G}(x, D) P^s_{x_0,\xi_0} \tag{4.12}$$

(product of the five operators: $Q_{x_0,\xi_0} I_s \mathcal{G}(x, D) I_{-s} P_{x_0,\xi_0}$)
appears as *a linear continuous mapping of $L^2(\mathbb{R}^n)$ into itself.*

5.

In this section we shall develop a representation formula for the action of operator $\mathcal{G}_s(x, D)$ on functions in $\mathcal{S}(\mathbb{R}^n)$ (the case $s = 0$ has been examined in Prop. 2.3).

First we have (see (1.7)), that the Fourier transform of $\mathcal{G}(x, D)v$, for $v \in \mathcal{S}$, is given by the expression:

$$(\mathcal{G}(x, D)v)^\wedge(\xi) = (2\pi)^{-n/2} \int_{\mathbb{R}^n} \gamma(\xi - \eta, \eta)\hat{v}(\eta)d\eta + \tilde{g}(\xi)\hat{v}(\xi), \xi \neq 0 \tag{5.1}$$

which is a function in $L^1_{loc}(\mathbb{R}^n_\xi)$.

In our present situation $v = P^s_{x_0,\xi_0} u$, where $u \in \mathcal{S}$, and (see (4.4)) $v \in \mathcal{S}$ too.

Next, the Fourier transform $\hat{v}$ is computed as follows: we have $v = I_{-s} P_{x_0,\xi_0} u = \mathcal{F}^{-1} \mathcal{M}_{(1+|\xi|^2)^{-s/2}} \mathcal{F} P_{x_0,\xi_0} u$; hence one obtains

$$\hat{v}(\xi) = \mathcal{M}_{(1+|\xi|^2)^{-s/2}} (P_{x_0,\xi_0} u)^\wedge(\xi) = (1 + |\xi|^2)^{-s/2} |\xi_0|^{-n/4} e^{-i<x_0,\xi-\xi_0>} \hat{u}(\frac{\xi - \xi_0}{\sqrt{|\xi_0|}}) \tag{5.2}$$

(using relation (2.10) above).

Consequently, using (5.1)–(5.2) we obtain by now:

$$(\mathcal{G}(x, D)P^s_{x_0,\xi_0} u)^\wedge(\xi) = \tilde{g}(\xi)(1 + |\xi|^2)^{-s/2} |\xi_0|^{-n/4} e^{-i<x_0,\xi-\xi_0>} \hat{u}(\frac{\xi - \xi_0}{\sqrt{|\xi_0|}})$$
$$+ (2\pi)^{-n/2} \int_{\mathbb{R}^n} \gamma(\xi - \eta, \eta)(1 + |\eta|^2)^{-s/2} |\xi_0|^{-n/4} e^{-i<x_0,\eta-\xi_0>} \hat{u}(\frac{\eta - \xi_0}{\sqrt{|\xi_0|}}) d\eta,$$
$$\forall \xi \in \mathbb{R}^n/\{o\} \tag{5.3}$$

Then we have to consider the expression $I_s \mathcal{G}(x, D) P^s_{x_0,\xi_0} u$; its Fourier transform appears to be equal to: $(1 + |\xi|^2)^{s/2} (\mathcal{G} P^s u)^\wedge(\xi)$, that is, from (5.3):

$$(I_s \mathcal{G}(x, D) P^s_{x_0,\xi_0} u)^\wedge(\xi) = \tilde{g}(\xi) |\xi_0|^{-n/4} e^{-i<x_0,\xi-\xi_0>} \hat{u}(\frac{\xi - \xi_0}{\sqrt{|\xi_0|}}) +$$

$$(2\pi)^{-n/2}(1+|\xi|^2)^{s/2}\int_{\mathbb{R}^n}\gamma(\xi-\eta,\eta)(1+|\eta|^2)^{-s/2}|\xi_0|^{-n/4}e^{-i<x_0,\eta-\xi_0>}\hat{u}(\frac{\eta-\xi_0}{\sqrt{|\xi_0|}})d\eta,$$
$$\forall\xi\in\mathbb{R}^n/\{o\} \quad (5.4)$$

The final step involves computation of the Fourier transformation

$$\mathcal{F}(Q_{x_0,\xi_0}I_s\mathcal{G}(x,D)P^s_{x_0,\xi_0}u) \quad (5.5)$$

We need a formula for $(Q_{x_0,\xi_0}w)^\wedge$, where $w\in L^2(\mathbb{R}^n)$.

(note that, if $u\in\mathcal{S}$, then $P^s_{x_0,\xi_0}u\in\mathcal{S},\mathcal{G}(x,D)P^s_{x_0,\xi_0}u\in H^s(\mathbb{R}^n)$, hence $I_s\mathcal{G}(x,D)P^s_{x_0,\xi_0}u\in L^2(\mathbb{R}^n)$).

Remember that

$$(Q_{x_0,\xi_0}w)(x)=|\xi_0|^{-n/4}w(x_0+|\xi_0|^{-1/2}x)e^{-i<x_0+|\xi_0|^{-1/2}x,\xi_0>},\forall w\in L^2(\mathbb{R}^n) \quad (5.6)$$

and that

$$\| Q_{x_0,\xi_0}w\|_{L^2}=\| w\|_{L^2}\ \forall w\in L^2 \quad (5.7)$$

Let us first compute the Fourier transform

$(Q_{x_0,\xi_0}w)^\wedge$ when $w\in\mathcal{S}(\mathbb{R}^n)$. In this case we simply have

$$(Q_{x_0,\xi_0}w)^\wedge(\xi)=(2\pi)^{-n/2}\int_{\mathbb{R}^n}e^{-i<x,\xi>}|\xi_0|^{-n/4}w(x_0+|\xi_0|^{-1/2}x)e^{-i<x_0+|\xi_0|^{-1/2}x,\xi_0>}dx$$

$=$ (after the substitution: $x_0+|\xi_0|^{-1/2}x=y,\ dx=|\xi_0|^{n/2}dy$)

$$(2\pi)^{-n/2}\int_{\mathbb{R}^n}e^{-i<|\xi_0|^{1/2}(y-x_0),\xi>}|\xi_0|^{-n/4}w(y)e^{-i<y,\xi_0>}|\xi_0|^{n/2}dy=$$

$$(2\pi)^{-n/2}|\xi_0|^{n/4}\int_{\mathbb{R}^n}e^{-i<y,|\xi_0|^{1/2}\xi+\xi_0>}e^{i<|\xi_0|^{1/2}x_0,\xi>}w(y)dy=$$

$$=|\xi_0|^{n/4}\hat{w}(\xi_0+|\xi_0|^{1/2}\xi)e^{i<x_0,|\xi_0|^{1/2}\xi>} \quad (5.8)$$

Let us define the operator R_{x_0,ξ_0} by the formula:

$$(R_{x_0,\xi_0}z)(\xi)=|\xi_0|^{n/4}z(\xi_0+|\xi_0|^{1/2}\xi)e^{i<x_0,|\xi_0|^{1/2}\xi>} \quad (5.9)$$

Note that R maps $\mathcal{S}(\mathbb{R}^n)$ into itself and also $L^2(\mathbb{R}^n)$ into itself. Moreover, R is isometric on $L^2(\mathbb{R}^n)$, as simply follows:

$$\| R_{x_0,\xi_0}z\|^2_{L^2}=\int_{\mathbb{R}^n}|\xi_0|^{n/2}|z(\xi_0+|\xi_0|^{1/2}\xi)|^2d\xi=(\text{making }\xi_0+|\xi_0|^{1/2}\xi=\eta)$$

$$\int_{\mathbb{R}^n}|\xi_0|^{n/2}|z(\eta)|^2|\xi_0|^{-n/2}d\eta=\| z\|^2_{L^2} \quad (5.10)$$

Now, previous calculations (5.8)–(5.9) show that

$$(Q_{x_0,\xi_0}w)^\wedge(\xi) = (R_{x_0,\xi_0}\hat{w})(\xi), \text{ for } w \in \mathcal{S}(\mathbb{R}^n), \xi \in \mathbb{R}^n \tag{5.11}$$

Take now any $w \in L^2$, and a sequence (w_p) in $\mathcal{S}, w_p \to w$ in L^2-sense and therefore $\hat{w}_p \to \hat{w}$ in L^2.

We know that $Q_{x_0,\xi_0}w \in L^2$ too, and its Fourier–Plancherel transform $\mathcal{F}(Q_{x_0,\xi_0}w)$ exists ($\in L^2$ too). Also, $\mathcal{F}(Q_{x_0,\xi_0}w) = \lim_{p\to\infty}\mathcal{F}(Q_{x_0,\xi_0}w_p)$ (in L^2-sense, as $\mathcal{F}$ is isometric in L^2).

$$\text{Hence: } \mathcal{F}(Q_{x_0,\xi_0}w) = \lim_{p\to\infty} R_{x_0,\xi_0}\hat{w}_p = R_{x_0,\xi_0}\hat{w}, \forall w \in L^2(\mathbb{R}^n) \tag{5.12}$$

(lastly we use: R is isometric in L^2 and $\hat{w}_p \to \hat{w}$ in L^2-sense).

Therefore we now have the formula (5.12), for the Fourier transform of $Q_{x_0,\xi_0}w$ when $w \in L^2(\mathbb{R}^n)$. We shall apply it taking $w = I_s\mathcal{G}(x,D)P^s_{x_0,\xi_0}u$; its Fourier transform $\hat{w}$ appears in (5.4):

$$\hat{w}(\xi) = \tilde{g}(\xi)|\xi_0|^{-n/4}e^{-i<x_0,\xi-\xi_0>}\hat{u}(\frac{\xi-\xi_0}{\sqrt{|\xi_0|}}) + (2\pi)^{-n/2}(1+|\xi|^2)^{s/2}$$
$$\int_{\mathbb{R}^n}\gamma(\xi-\eta,\eta)(1+|\eta|^2)^{-s/2}|\xi_0|^{-n/4}e^{-i<x_0,\eta-\xi_0>}\hat{u}(\frac{\eta-\xi_0}{\sqrt{|\xi_0|}})d\eta, \xi \neq 0$$

and therefore:

$$(R_{x_0,\xi_0}\hat{w})(\xi) = |\xi_0|^{n/4}\hat{w}(\xi_0+|\xi_0|^{1/2}\xi)e^{i<x_0,|\xi_0|^{1/2}\xi>} =$$
$$|\xi_0|^{n/4}\tilde{g}(\xi_0+|\xi_0|^{1/2}\xi)|\xi_0|^{-n/4}e^{-i<x_0,\xi|\xi_0|^{1/2}+\xi_0-\xi_0>}\hat{u}(\frac{\xi|\xi_0|^{1/2}+\xi_0-\xi_0}{\sqrt{|\xi_0|}})\cdot e^{i<x_0,|\xi_0|^{1/2}\xi>}$$
$$+(2\pi)^{-n/2}|\xi_0|^{n/4}e^{i<x_0,|\xi_0|^{1/2}\xi>}(1+|\xi_0+|\xi_0|^{1/2}\xi|^2)^{s/2}. \tag{5.13}$$
$$\int_{\mathbb{R}^n}\gamma(\xi|\xi_0|^{1/2}+\xi_0-\eta,\eta)(1+|\eta|^2)^{-s/2}|\xi_0|^{-n/4}e^{-i<x_0,\eta-\xi_0>}\hat{u}(\frac{\eta-\xi_0}{\sqrt{|\xi_0|}})d\eta$$
$$= A(\xi)+B(\xi),\ \xi \neq -|\xi_0|^{-1/2}\xi_0$$

Actually we see that

$$A(\xi) = \tilde{g}(\xi_0+|\xi_0|^{1/2}\xi)\cdot\hat{u}(\xi), \xi \neq -|\xi_0|^{-1/2}\xi_0 \tag{5.14}$$
$$B(\xi) = \left((2\pi)^{-n/2}\int_{\mathbb{R}^n}\gamma(\xi|\xi_0|^{1/2}+\xi_0-\eta,\eta)(1+|\eta|^2)^{-s/2}e^{-i<x_0,\eta-\xi_0>}\hat{u}(\frac{\eta-\xi_0}{\sqrt{|\xi_0|}}d\eta\right)\cdot$$

$$e^{i<x_0,|\xi_0|^{1/2}\xi>}(1+|\xi_0+\xi|\xi_0|^{1/2}|^2)^{s/2} = (\text{after substitution } \xi\sqrt{|\xi_0|}+\xi_0-\eta=\sigma).$$

$$\Big((2\pi)^{-n/2}\int_{\mathbb{R}^n}\gamma(\sigma,\xi\sqrt{|\xi_0|}+\xi_0-\sigma)(1+|\xi\sqrt{|\xi_0|}+\xi_0-\sigma|^2)^{-s/2}$$

$$e^{-i<x_0,\xi\sqrt{|\xi_0|}-\sigma>}\hat{u}(\xi-\frac{\sigma}{\sqrt{|\xi_0|}})d\sigma\Big)\cdot(e^{i<x_0,\xi\sqrt{|\xi_0|}>}(1+|\xi\sqrt{|\xi_0|}+\xi_0|^2)^{s/2}) =$$

$$(2\pi)^{-n/2}\int_{\mathbb{R}^n}\gamma(\sigma,\xi\sqrt{|\xi_0|}+\xi_0-\sigma)\Big(\frac{1+|\xi\sqrt{|\xi_0|}+\xi_0|^2}{1+|\xi\sqrt{|\xi_0|}+\xi_0-\sigma|^2}\Big)^{s/2}$$

$$e^{i<x_0,\sigma>}\hat{u}(\xi-\frac{\sigma}{\sqrt{|\xi_0|}})d\sigma \tag{5.15}$$

We can say accordingly that the Fourier transform

$(\mathcal{G}_s(x,D)u)^\wedge(\xi) = (Q_{x_0,\xi_0}I_s\mathcal{G}(x,D)P^s_{x_0,\xi_0}u)^\wedge(\xi)$ (for $u\in\mathcal{S}(\mathbb{R}^n)$), equals

$$\tilde{g}(\xi_0+\sqrt{|\xi_0|}\xi)\hat{u}(\xi)+(2\pi)^{-n/2}\int_{\mathbb{R}^n}\gamma(\sigma,\xi\sqrt{|\xi_0|}+\xi_0-\sigma)\Big(\frac{1+|\xi\sqrt{|\xi_0|}+\xi_0|^2}{1+|\xi\sqrt{|\xi_0|}+\xi_0-\sigma|^2}\Big)^{s/2}$$

$$e^{i<x_0,\sigma>}\hat{u}(\xi-\frac{\sigma}{\sqrt{|\xi_0|}})d\sigma \text{ for } \xi\neq-|\xi_0|^{-1/2}\xi_0 \tag{5.16}$$

This is the representation formula that we were looking for.

Remark Let us note the estimates:

$$|\gamma(\sigma,\xi\sqrt{|\xi_0|}+\xi_0-\sigma)|\le k(\sigma) \qquad \text{(from (1.2))}$$

$$\Big(\frac{1+|\xi\sqrt{|\xi_0|}+\xi_0|^2}{1+|\xi\sqrt{|\xi_0|}+\xi_0-\sigma|^2}\Big)^{s/2}\le 2^{\frac{|s|}{2}}(1+|\sigma|^2)^{\frac{|s|}{2}} \qquad \text{(Peetre's inequality, see [9])}$$

$$|\hat{u}(\xi-\frac{\sigma}{\sqrt{|\xi_0|}})|\le M \qquad (\text{as } \hat{u}\in\mathcal{S}(\mathbb{R}^n)).$$

Then, the term under integral sign in (5.16) is estimated, in absolute value, by: $2^{|s|/2}M(1+|\sigma|^2)^{|s|/2}k(\sigma)$, which belongs to $L^1(\mathbb{R}^n)$ by (1.3).

6.

We are now ready to establish following

Theorem 6.1 *Let be* $g(x,\xi),\mathbb{R}^n\times\mathbb{R}^n/\{o\}\to\mathbb{C}$ *a symbol satisfying* (1.1)–(1.2)–(1.3)–(1.4)– (1.5)–(1.6).

Assume that, for some $x_0\in\mathbb{R}^n$ *and a sequence* (ξ_p) *in* $\mathbb{R}^n/\{o\}$*, with* $\lim_{p\to\infty}|\xi_p|=+\infty$*, we have:* $\lim_{p\to\infty}g(x_0,\xi_p)=c_0\in\mathbb{C}$.

Then, there is a subsequence $(\xi_{p_k})_{k=1}^{\infty}$ *so that*

$$\lim_{k\to\infty}(Q^s_{x_0,\xi_{p_k}}\mathcal{G}(x,D)P^s_{x_0,\xi_{p_k}}u)^\wedge(\xi)=c_0\hat{u}(\xi) \tag{6.1}$$

holds, $\forall u\in\mathcal{S}(\mathbb{R}^n),\forall\xi\in\mathbb{R}^n$.

Proof We write: $g(x_0,\xi)=\tilde{g}(\xi)+(2\pi)^{-n/2}\int_{\mathbb{R}^n}e^{i<x_0,\lambda>}\gamma(\lambda,\xi)d\lambda,\xi\neq 0$. As $\tilde{g}(\xi)$ is bounded, there is a subsequence (ξ_{p_k}) such that $\tilde{g}(\xi_{p_k})\to b_0$ (in $\mathbb{C}$). It follows that $g(x_0,\xi_{p_k})\to c_0$ and then $g_1(x_0,\xi_{p_k})=(2\pi)^{-n/2}\int_{\mathbb{R}^n}e^{i<x_0,\lambda>}\gamma(\lambda,\xi_{p_k})d\lambda$ is convergent to $c_0-b_0=d_0\in\mathbb{C}$.

Next, we shall use (5.16); we first note that (for $k\in\mathbb{N}$ such that $\sqrt{|\xi_{p_k}|}>|\xi|$),

$$\lim_{k\to\infty}\tilde{g}(\xi_{p_k}+\sqrt{|\xi_{p_k}|}\xi)=b_0,\ \forall\xi\in\mathbb{R}^n \tag{6.2}$$

(use inequality: $|\xi_{p_k}+\sqrt{|\xi_{p_k}|}\xi|\geq\sqrt{|\xi_{p_k}|}(\sqrt{|\xi_{p_k}|}-|\xi|)$)

In fact, we have the simple estimates, for $k>k^-(\xi)$,

$$\begin{aligned}|\tilde{g}(\xi_{p_k}+\sqrt{|\xi_{p_k}|}\xi)-b_0|&\leq|\tilde{g}(\xi_{p_k}+\sqrt{|\xi_{p_k}|}\xi)-\tilde{g}(\xi_{p_k})|+|\tilde{g}(\xi_{p_k})-b_0|\\&\leq(1+\sqrt{|\xi_{p_k}|}|\xi|)\varphi(|\xi_{p_k}|)+|\tilde{g}(\xi_{p_k})-b_0|\end{aligned} \tag{6.3}$$

(by use of assumption (1.6)).

Finally, the expression in (6.3) is convergent to $0\forall\xi\in\mathbb{R}^n$, as follows from the assumption $\sqrt{t}\varphi(t)\to 0$ as $t\to+\infty$.

Next, we shall consider the limit

$$\lim_{p\to\infty}(2\pi)^{-n/2}\int_{\mathbb{R}^n}\gamma(\sigma,\xi\sqrt{|\xi_p|}+\xi_p-\sigma)\left(\frac{1+|\xi_p+\sqrt{|\xi_p|}\xi|^2}{1+|\xi_p-\sigma+\sqrt{|\xi_p|}\xi|^2}\right)^{s/2} e^{i<x_0,\sigma>}\hat{u}(\xi-\frac{\sigma}{\sqrt{|\xi_p|}})d\sigma \tag{6.4}$$

(precisely, we shall find a subsequence $(\xi_{p_{k_j}})_1^\infty$ of $(\xi_{p_k})_{k=1}^\infty$ such that the corresponding limit in (6.4) is c_0-b_0).

To start with, let us note the simple relation:

$$\frac{1+|\xi_p+\sqrt{|\xi_p|}\xi|^2}{1+|\xi_p-\sigma+\xi\sqrt{|\xi_p|}|^2}=\frac{1+|\xi_p|^2|\frac{\xi_p}{|\xi_p|}+\frac{\xi}{\sqrt{|\xi_p|}}|^2}{1+|\xi_p|^2|\frac{\xi_p}{|\xi_p|}-\frac{\sigma}{|\xi_p|}+\frac{\xi}{\sqrt{|\xi_p|}}|^2}=\frac{\frac{1}{|\xi_p|^2}+|\frac{\xi_p}{|\xi_p|}+\frac{\xi}{\sqrt{|\xi_p|}}|^2}{\frac{1}{|\xi_p|^2}+|\frac{\xi_p}{|\xi_p|}-\frac{\sigma}{|\xi_p|}+\frac{\xi}{\sqrt{|\xi_p|}}|^2} \tag{6.5}$$

(In fact, we write (6.5) for the subsequence (ξ_{p_k}) in (6.2)).

Note also that $|\xi_{p_k}| \to +\infty$ as $k \to \infty$ and that the vectors $\frac{\xi_{p_k}}{|\xi_{p_k}|}$ are situated on the unit sphere in $\mathbb{R}^n$; there is accordingly (another) subsequence $(\xi_{p_{k_j}})_{j=1}^{\infty}$ such that $\frac{\xi_{p_{k_j}}}{|\xi_{p_{k_j}}|} \to \xi_0$ as $j \to \infty$, where $|\xi_0| = 1$.

Looking now at (6.5) for this subsequence $(\xi_{p_{k_j}})_{j=1}^{\infty}$, we see that

$$\lim_{j\to\infty} \frac{\frac{1}{|\xi_{p_{k_j}}|^2} + |\frac{\xi_{p_{k_j}}}{|\xi_{p_{k_j}}|} + \frac{\xi}{\sqrt{|\xi_{p_{k_j}}|}}|^2}{\frac{1}{|\xi_{p_{k_j}}|^2} + |\frac{\xi_{p_{k_j}}}{|\xi_{p_{k_j}}|} - \frac{\sigma}{|\xi_{p_{k_j}}|} + \frac{\xi}{\sqrt{|\xi_{p_{k_j}}|}}|^2} = \frac{|\xi_0|^2}{|\xi_0|^2} = 1, \forall \xi \in \mathbb{R}^n, \forall \sigma \in \mathbb{R}^n \qquad (6.6)$$

We now turn back to (6.4) and write the following formula:

$$(2\pi)^{-n/2} \int_{\mathbb{R}^n} \gamma(\sigma, \xi\sqrt{|\xi_{p_{k_j}}|} + \xi_{p_{k_j}} - \sigma)\left(\frac{1 + |\xi_{p_{k_j}} + \sqrt{|\xi_{p_{k_j}}|}\xi|^2}{1 + |\xi_{p_{k_j}} - \sigma + \sqrt{|\xi_{p_{k_j}}|}\xi|^2}\right)^{s/2}$$

$$e^{i<x_0,\sigma>}\hat{u}(\xi - \frac{\sigma}{\sqrt{|\xi_{p_{k_j}}|}})d\sigma = \mathcal{J}_j(\xi) + L_j(\xi). \qquad (6.7)$$

where we denote:

$$\mathcal{J}_j(\xi) = (2\pi)^{-n/2} \int_{\mathbb{R}^n} [\gamma(\sigma, \xi\sqrt{|\xi_{p_{k_j}}|} + \xi_{p_{k_j}} - \sigma) - \gamma(\sigma, \xi_{p_{k_j}})].$$

$$\left(\frac{1 + |\xi_{p_{k_j}} + \sqrt{|\xi_{p_{k_j}}|}\xi|^2}{1 + |\xi_{p_{k_j}} - \sigma + \sqrt{|\xi_{p_{k_j}}|}\xi|^2}\right)^{s/2} e^{i<x_0,\sigma>}\hat{u}(\xi - \frac{\sigma}{\sqrt{|\xi_{p_{k_j}}|}})d\sigma \qquad (6.8)$$

and

$$L_j(\xi) = (2\pi)^{-n/2} \int_{\mathbb{R}^n} \gamma(\sigma, \xi_{p_{k_j}})\left(\frac{1 + |\xi_{p_{k_j}} + \sqrt{|\xi_{p_{k_j}}|}\xi|^2}{1 + |\xi_{p_{k_j}} - \sigma + \sqrt{|\xi_{p_{k_j}}|}\xi|^2}\right)^{s/2}$$

$$e^{i<x_0,\sigma>}\hat{u}(\xi - \frac{\sigma}{\sqrt{|\xi_{p_{k_j}}|}})d\sigma \qquad (6.9)$$

Note also the further (obvious) decomposition

$$L_j(\xi) = (2\pi)^{-n/2} \int_{\mathbb{R}^n} \gamma(\sigma, \xi_{p_{k_j}})e^{i<x_0,\sigma>}\hat{u}(\xi)d\sigma +$$

$$(2\pi)^{-n/2}\int_{\mathbb{R}^n}\gamma(\sigma,\xi_{p_{k_j}})e^{i<x_0,\sigma>}\Bigg[\left(\frac{1+|\xi_{p_{k_j}}+\sqrt{|\xi_{p_{k_j}}|}\xi|^2}{1+|\xi_{p_{k_j}}-\sigma+\sqrt{|\xi_{p_{k_j}}|}\xi|^2}\right)^{s/2}$$

$$\hat{u}(\xi-\frac{\sigma}{\sqrt{|\xi_{p_{k_j}}|}})-\hat{u}(\xi)\Bigg]d\sigma = M_j(\xi)+N_j(\xi) \tag{6.10}$$

Actually, we see that (using (1.1)):

$$M_j(\xi)=\hat{u}(\xi)[g(x_0,\xi_{p_{k_j}})-\tilde{g}(\xi_{p_{k_j}})] \tag{6.11}$$

Next, we make appeal to the limits:
$\tilde{g}(\xi_{p_{k_j}})\to b_0$ and $g(x_0,\xi_{p_{k_j}})\to c_0$ (as $j\to\infty$) and we see that

$$\lim_{j\to\infty} M_j(\xi)=(c_0-b_0)\hat{u}(\xi), \forall\xi\in\mathbb{R}^n. \tag{6.12}$$

We also establish that:

$$\lim_{j\to\infty}\mathcal{J}_j(\xi)=\lim_{j\to\infty}N_j(\xi)=0,\ \forall\xi\in\mathbb{R}^n \tag{6.13}$$

In order to get the first limit, we start with the (simple) estimate

$$|\mathcal{J}_j(\xi)|\le(2\pi)^{-n/2}\int_{\mathbb{R}^n}(1+|\xi\sqrt{|\xi_{p_{k_j}}|}-\sigma|)\varphi(|\xi_{p_{k_j}}|)k(\sigma).2^{|s|/2}$$

$$(1+|\sigma|^2)^{|s|/2}|\hat{u}(\xi-\frac{\sigma}{\sqrt{|\xi_{p_{k_j}}|}})|d\sigma \tag{6.14}$$

which is obtained using (1.5) for

$$\sigma\ne\xi\sqrt{|\xi_{p_{k_j}}|}+\xi_{p_{k_j}}, \text{ and Peetre's inequality.}$$

Denote also: $\xi\sqrt{|\xi_{p_{k_j}}|}-\sigma=\mu\sqrt{|\xi_{p_{k_j}}|}$; we get the inequality

$$1+|\xi\sqrt{|\xi_{p_{k_j}}|}-\sigma|=1+|\mu|\sqrt{|\xi_{p_{k_j}}|}\le\sqrt{|\xi_{p_{k_j}}|}(1+|\mu|) \tag{6.15}$$

for sufficiently large j (as $|\xi_{p_{k_j}}|\to\infty$!).

Accordingly, we obtain the estimate:

$$(1+|\xi\sqrt{|\xi_{p_{k_j}}|}-\sigma||\hat{u}(\xi-\frac{\sigma}{\sqrt{|\xi_{p_{k_j}}|}})|=$$

$$1+|\mu|\sqrt{|\xi_{p_{k_j}}|}|\hat{u}(\mu)|\leq(1+|\mu|)\sqrt{|\xi_{p_{k_j}}|}|\hat{u}(\mu)|$$

$$\leq M\sqrt{|\xi_{p_{k_j}}|} \text{ where } M=\sup_{\eta\in\mathbb{R}^n}(1+|\eta|)|\hat{u}(\eta)|<\infty\ (\hat{u}\in\mathcal{S}!) \tag{6.16}$$

Introducing (6.16) in (6.14) we now have:

$$|\mathcal{J}_j(\xi)|\leq(2\pi)^{-n/2}M\sqrt{|\xi_{p_{k_j}}|}\varphi(|\xi_{p_{k_j}}|).2^{|s|/2}\int_{\mathbb{R}^n}(1+|\sigma|^2)^{|s|/2}k(\sigma)d\sigma$$
$$=C_s\sqrt{|\xi_{p_{k_j}}|}\varphi(|\xi_{p_{k_j}}|) \tag{6.17}$$

which $\to 0$ as $j\to\infty$, due to:

$\sqrt{t}\varphi(t)\to 0$ as $t\to\infty$.

We now establish the second limit in (6.13): $\lim_{j\to\infty} N_j(\xi)=0\ \forall\xi\in\mathbb{R}^n$.

We have (see (6.10)):

$$N_j(\xi)=(2\pi)^{-n/2}\int_{\mathbb{R}^n}\gamma(\sigma,\xi_{p_{k_j}})e^{i<x_0,\sigma>}\left[\left(\frac{1+|\xi_{p_{k_j}}+\sqrt{|\xi_{p_{k_j}}|}\xi|^2}{1+|\xi_{p_{k_j}}-\sigma+\sqrt{|\xi_{p_{k_j}}|}.\xi|^2}\right)^{s/2}\right.$$
$$\left.\hat{u}(\xi-\frac{\sigma}{\sqrt{|\xi_{p_{k_j}}|}})-\hat{u}(\xi)\right]d\sigma;$$

we obtain the estimate for the term under integral sign: it is $\leq$

$$k(\sigma)\left[2^{|s|/2}(1+|\sigma|^2)^{|s|/2}|\hat{u}(\xi-\frac{\sigma}{\sqrt{|\xi_{p_{k_j}}|}})|+|\hat{u}(\xi)|\right]\leq$$

$$2C_sM(1+|\sigma|^2)^{|s|/2}k(\sigma),\text{ which belongs to } L^1(\mathbb{R}^n) \tag{6.18}$$
$$(\text{where } M=\sup_{\tau\in\mathbb{R}^n}|\hat{u}(\tau)|)$$

Furthermore, we can also estimate differently the term under the integral sign by

$$k(\sigma)\left|\left[\left(\frac{1+|\xi_{p_{k_j}}+\xi\sqrt{|\xi_{p_{k_j}}|}|^2}{1+|\xi_{p_{k_j}}-\sigma+\sqrt{|\xi_{p_{k_j}}|}.\xi|^2}\right)^{s/2}\hat{u}(\xi-\frac{\sigma}{\sqrt{|\xi_{p_{k_j}}|}})-\hat{u}(\xi)\right]\right|, \text{ for } \xi\in\mathbb{R}^n. \tag{6.19}$$

If we fix $\sigma \in \mathbb{R}^n$, we see that: $\frac{1+|\xi_{p_{k_j}}+\xi\sqrt{|\xi_{p_{k_j}}|}|^2}{1+|\xi_{p_{k_j}}-\sigma+\sqrt{|\xi_{p_{k_j}}|}.\xi|^2} \to 1$(as $j \to \infty$ (from (6.5)–(6.6)) and that $\hat{u}(\xi - \frac{\sigma}{\sqrt{|\xi_{p_{k_j}}|}}) \to \hat{u}(\xi)$ (as $j \to \infty$) (this, $\forall\xi \in \mathbb{R}^n$).

Therefore, the expression (6.19) $\to 0 \forall\sigma \in \mathbb{R}^n$ where $k(\sigma) < \infty$ (hence a.e. on $\mathbb{R}^n$).

Applying the dominated convergence theorem we find : $\lim_{j\to\infty} N_j(\xi) = 0, \forall\xi \in \mathbb{R}^n$.

Thus, putting all above together we have found that, $\forall u \in \mathcal{S}(\mathbb{R}^n), \forall\xi \in \mathbb{R}^n$

$$(Q^s_{x_0,\xi_{p_{k_j}}}\mathcal{G}(x,D)P^s_{x_0,\xi_{p_{k_j}}}u)^\wedge(\xi) = \tilde{g}(\xi_{p_{k_j}}+\sqrt{|\xi_{p_{k_j}}|}\xi)\hat{u}(\xi)+$$

$$(2\pi)^{-n/2}\int_{\mathbb{R}^n}\gamma(\sigma,\xi\sqrt{|\xi_{p_{k_j}}|}+\xi_{p_{k_j}}-\sigma)\left(\frac{1+|\xi\sqrt{|\xi_{p_{k_j}}|}+\xi_{p_{k_j}}|^2}{1+|\xi\sqrt{|\xi_{p_{k_j}}|}+\xi_{p_{k_j}}-\sigma|^2}\right)^{s/2}$$

$$e^{i<x_0,\sigma>}\hat{u}(\xi-\frac{\sigma}{\sqrt{|\xi_{p_{k_j}}|}})d\sigma = \tilde{g}(\xi_{p_{k_j}}+\sqrt{|\xi_{p_{k_j}}|}\xi)\hat{u}(\xi)+\mathcal{J}_j(\xi)+M_j(\xi)+N_j(\xi) \tag{6.20}$$

Using relations (6.2), (6.12), (6.13) we obtain the final result:

$$\lim_{j\to\infty}(Q^s_{x_0,\xi_{p_{k_j}}}\mathcal{G}(x,D)P^s_{x_0,\xi_{p_{k_j}}}u)^\wedge(\xi) = b_0\hat{u}(\xi)+(c_0-b_0)\hat{u}(\xi) = c_0\hat{u}(\xi),\ \forall\xi\in\mathbb{R}^n$$

which is (6.1).

This establishes Th. 6.1.

7.

In this section our main result is the following

Theorem 7.1 *Under the assumptions of Th. 6.1 we have the relation*

$$\lim_{j\to\infty} Q^s_{x_0,\xi_{p_{k_j}}}\mathcal{G}(x,D)P^s_{x_0,\xi_{p_{k_j}}}u = c_0 u \tag{7.1}$$

in weak $L^2(\mathbb{R}^n)$*-sense,* $\forall u \in L^2(\mathbb{R}^n)$

(where $\xi_{p_{k_j}}$ is the subsubsequence of (ξ_p) which appears in the proof of Th. 6.1).

Remark 1 As seen in section 4, the operator $\mathcal{G}_s(x, D)$ given by (4.12) is a linear continuous mapping, $L^2(\mathbb{R}^n) \to L^2(\mathbb{R}^n)$.

Remark 2 Let us define a sequence $(v_j)_1^\infty$ where $v_j = \mathcal{G}_{s,j}u$, with $\mathcal{G}_{s,j}$ the operator in the left-hand side of (7.1). We see that

$$\begin{aligned} \| v_j \|_{L^2} &= \| Q^s_{x_0,\xi_{p_{k_j}}} (\mathcal{G}(x,D) P^s_{x_0,\xi_{p_{k_j}}} u) \|_{L^2} = \| \mathcal{G}(x,D) P^s_{x_0,\xi_{p_{k_j}}} u \|_{H^s} \\ &\leq \| \mathcal{G}(x,D) \|_{\mathcal{L}(H^s,H^s)} \cdot \| P^s_{x_0,\xi_{p_{k_j}}} u \|_{H^s} = \| \mathcal{G}(x,D) \|_{\mathcal{L}(H^s)} \| u \|_{L^2} \end{aligned} \tag{7.2}$$

so that $(v_j)_1^\infty$ is bounded in $L^2(\mathbb{R}^n)$.

Proof of Th. 7.1

A) First we establish the relation (7.1) for all $u \in \mathcal{S}(\mathbb{R}^n)$. The weak convergence means, as usual, that the following holds:

$$\lim_{j\to\infty} \int_{\mathbb{R}^n} (\mathcal{G}_{s,j}u)(x)\bar{h}(x)dx = \int_{\mathbb{R}^n} c_0 u(x)\bar{h}(x)dx \ \forall u \in \mathcal{S},\ \forall h \in L^2(\mathbb{R}^n) \tag{7.3}$$

Using Parseval–Plancherel theorem, we see that (7.3) means also that

$$\lim_{j\to\infty} \int_{\mathbb{R}^n} (\mathcal{G}_{s,j}u)^\wedge(\xi)\bar{\hat{h}}(\xi)d\xi = c_0 \int_{\mathbb{R}^n} \hat{u}(\xi)\bar{\hat{h}}(\xi)d\xi \ \forall u \in \mathcal{S},\ \forall h \in L^2(\mathbb{R}^n) \tag{7.4}$$

Thus, let us establish (7.4) for $u \in \mathcal{S}(\mathbb{R}^n)$, In this case the Fourier transform $(\mathcal{G}_{s,j}u)^\wedge(\xi)$ is expressed by: $\tilde{g}(\xi_{p_{k_j}} + \sqrt{|\xi_{p_{k_j}}|}\xi)\hat{u}(\xi) +$

$$(2\pi)^{-n/2} \int_{\mathbb{R}^n} \gamma(\sigma, \xi\sqrt{|\xi_{p_{k_j}}|} + \xi_{p_{k_j}} - \sigma)\left(\frac{1 + |\xi\sqrt{|\xi_{p_{k_j}}|} + \xi_{p_{k_j}}|^2}{1 + |\xi\sqrt{|\xi_{p_{k_j}}|} + \xi_{p_{k_j}} - \sigma|^2}\right)^{s/2}$$

$$e^{i<x_0,\sigma>}\hat{u}(\xi - \frac{\sigma}{\sqrt{|\xi_{p_{k_j}}|}})d\sigma.$$

and we know already that $(\mathcal{G}_{s,j}u)^\wedge(\xi) \to c_0\hat{u}(\xi),\ \forall \xi \in \mathbb{R}^n$.

Remark 3. If we could find an estimate of the form

$$|(\mathcal{G}_{s,j}u)^\wedge(\xi)\bar{\hat{h}}(\xi)| \leq H(\xi) \ \text{ where } H \in L^1(\mathbb{R}^n)$$

we could derive immediately (7.4) from the dominated convergence theorem.

Trying a simple estimate we obtain:

$$|(\mathcal{G}_{s,j}u)^\wedge(\xi)| \leq C|\hat{u}(\xi)| + (2\pi)^{-n/2}\int_{\mathbb{R}^n} k(\sigma)2^{|s|/2}(1+|\sigma|^2)^{|s|/2}|\hat{u}(\xi - \frac{\sigma}{\sqrt{|\xi_{p_{k_j}}|}})|d\sigma$$

$$\leq C|\hat{u}(\xi)| + (2\pi)^{-n/2}\sup_{\tau\in\mathbb{R}^n}|\hat{u}(\tau)|2^{|s|/2}\int_{\mathbb{R}^n}(1+|\sigma|^2)^{|s|/2}k(\sigma)d\sigma = C|\hat{u}(\xi)| + C_1.$$

Then $|(\mathcal{G}_{s,j}u)^\wedge(\xi)\bar{\hat{h}}(\xi)| \leq C|\hat{u}(\xi)||\hat{h}(\xi)| + C_1|\hat{h}(\xi)|$;
as $|\hat{h}| \in L^2(\mathbb{R}^n)$ but not necessarily in $L^1(\mathbb{R}^n)$, we don't get the good inequality.

Thus, let us *prove* (7.4) for all $u \in \mathcal{S}(\mathbb{R}^n)$; we shall reason by contradiction assuming that, for at least one $h \in L^2(\mathbb{R}^n)$, the limit (7.4) does *not* hold (for some $u \in \mathcal{S}(\mathbb{R}^n)$ which is now fixed!).

Hence, the numerical sequence (for this couple u, h):

$$\alpha_j = \int_{\mathbb{R}^n}(\mathcal{G}_{s,j}u)^\wedge(\xi)\bar{\hat{h}}(\xi)d\xi \text{ is } \textit{not} \text{ convergent to } c_0\int_{\mathbb{R}^n}\hat{u}(\xi)\bar{\hat{h}}(\xi)d\xi$$

As a corollary of this fact, we see that there is an $\mathcal{E} > 0$ and a subsequence $(\alpha_{j_k})_1^\infty$ such that

$$|\alpha_{j_k} - c_0\int_{\mathbb{R}^n}\hat{u}(\xi)\bar{\hat{h}}(\xi)d\xi| > \mathcal{E}\ \forall k = 1,2,3,\ldots \tag{7.5}$$

Let us use now Remark 2. Therefore, the sequence $(\mathcal{G}_{s,j_k}u)^\wedge$ is also $L^2(\mathbb{R}^n)$-bounded. By classical properties, we can extract a further subsequence $((\mathcal{G}_{s,j_{k_\ell}}u)^\wedge(\xi))_{\ell=1}^\infty$ which is weakly-L^2 convergent towards $w \in L^2(\mathbb{R}^n)$.

Thus, we now have the situation where

$$\begin{gathered}(\mathcal{G}_{s,j_{k_\ell}}u)^\wedge(\xi) \to c_0\hat{u}(\xi) \text{ pointwise in } \mathbb{R}^n \\ \text{and in same time} \\ (\mathcal{G}_{s,j_{k_\ell}}u)^\wedge(\xi) \to w(\xi) \text{ weakly in } L^2(\mathbb{R}^n)\end{gathered} \tag{7.6}$$

We see that from (7.6) it follows: $c_0\hat{u}(\xi) = w(\xi)$ a.e. in $\mathbb{R}^n$.

In fact, let $\psi_\ell(\xi) = (\mathcal{G}_{s,j_{k_\ell}}u)^\wedge(\xi) - w(\xi)$: then: $\psi_\ell \to 0$ weakly in L^2 and $\psi_\ell(\xi) \to c_0\hat{u}(\xi) - w(\xi), \forall\xi \in \mathbb{R}^n$.

It will follow that: $c_0\hat{u}(\xi) - w(\xi) = 0$, a.e. in $\mathbb{R}^n$. In fact, we establish the following (general)

Proposition *Let $(\phi_\ell(\xi))_1^\infty$ be a sequence of $L^2(\mathbb{R}^n)$-functions, such that*

i) $\phi_\ell \to 0$ *weakly in* L^2 (7.7)

ii) $\phi_\ell(\xi) \to \phi(\xi) \in L^2(\mathbb{R}^n)$, *a.e. on* $\mathbb{R}^n$. (7.8)

Then $\phi(\xi) = 0$ *a.e.*

Proof Take $\mathbb{R}_e\phi_\ell$ and $\mathcal{J}_m\phi_\ell$. Then $\mathbb{R}_e\phi_\ell(\xi) \to \mathbb{R}_e\phi(\xi)$, a.e. on $\mathbb{R}^n$, and (as $\mathbb{R}_e\phi_\ell = \frac{1}{2}(\phi_\ell + \bar{\phi}_\ell))$ we also get $\mathbb{R}_e\phi_\ell \to 0$ weakly in L^2; (note that: $\phi_\ell \to$ (weakly in L^2)$\phi \Rightarrow \bar{\phi}_\ell \to$ (weakly in L^2) $\bar{\phi}$, as readily seen). In similar way: $\mathcal{J}_m\phi_\ell(\xi) \to \mathcal{J}_m\phi(\xi)$ a.e. on $\mathbb{R}^n, \mathcal{J}_m\phi_\ell(\xi) \to 0$ weakly in $L^2(\mathbb{R}^n)$.

It is therefore sufficient to prove the *Proposition* for real-valued functions only. We shall proceed by contradiction:

assume therefore that $m\{\xi;\phi(\xi) \neq 0\} > 0$. Denote $\mathcal{E}_p = \{\xi \in \mathbb{R}^n, |\phi(\xi)| > \frac{1}{p}\}$. Then we see that $\{\xi;\phi(\xi) \neq 0\} = \bigcup_{p=1}^\infty \mathcal{E}_p$ and $m(\mathcal{E}_p) = 0 \forall p = 1,2,\ldots$ is impossible. Thus, for some $p_0 \in \mathbb{N}$, we have $m(\mathcal{E}_{p_0}) > 0$ and $|\phi(\xi)| > \frac{1}{p_0}$ for $\xi \in \mathcal{E}_{p_0}$. If $\mathcal{E}_{p_0}^+ = \{\xi, \phi(\xi) > \frac{1}{p_0}\}$ and $\mathcal{E}_{p_0}^- = \{\xi, \phi(\xi) < -\frac{1}{p_0}\}$ we see that $\mathcal{E}_{p_0} = \mathcal{E}_{p_0}^+ \cup \mathcal{E}_{p_0}^-$ and at least one of these 2 sets has positive measure.

Let, say, be $m(\mathcal{E}_{p_0}^+) > 0$.

Next, let us use ii); it results that

$$\phi_n(\xi) \to \phi(\xi) \text{ a.e. on } \mathcal{E}_{p_0}^+ (\text{hence } \forall\xi \in \mathcal{E}_{p_0}^+/S_0,\ m(S_0) = 0).$$

Consequently, we obtain:

$$\forall\xi \in \mathcal{E}_{p_0}^+/S_0, \exists\bar{j} \in \mathbb{N},\ \bar{j} = \bar{j}(\xi), \text{ such that: } \phi_j(\xi) > \frac{1}{p_0} \text{ for } j \geq \bar{j}$$

Define now the set

$$S_k = \{\xi \in \mathcal{E}_{p_0}^+/S_0;\ \bar{j}(\xi) = k\}.$$

It follows: $\mathcal{E}_{p_0}^+/S_0 = \bigcup_{k=1}^\infty S_k$. (In fact, each S_k is contained in $\mathcal{E}_{p_0}^+/S_0$; this gives $\bigcup_1^\infty S_k \subset \mathcal{E}_{p_0}^+/S_0$; next, each $\xi \in \mathcal{E}_{p_0}^+/S_0$ belongs to some $S_{\bar{j}}$, hence we have the other inclusion).

Now: $m(\mathcal{E}_{p_0}^+/S_0) > 0$ – as $m(\mathcal{E}_{p_0}^+) > 0$ and $m(S_0) = 0$.

It results that: $m(S_{\bar{k}}) > 0$ for at least one $\bar{k} \in \mathbb{N}$ and we have, from definition of $S_{\bar{k}}$, that

$$\phi_j(\xi) > \frac{1}{p_0} \text{ for } j \geq \bar{k},\ \forall\xi \in S_{\bar{k}}/S_0$$

Define now the set: $S_{\bar{k}}^{\ell} = \{\xi \in S_{\bar{k}}, |\xi| \leq \ell\}$ (where $\ell \in \mathbb{N}$).

Again we have

$S_{\bar{k}} = \bigcup_{\ell=1}^{\infty} S_{\bar{k}}^{\ell}$, hence at least one $S_{\bar{k}}^{\bar{\ell}}$ has positive measure.

Let us define: $g(\xi) = 1$ on $S_{\bar{k}}^{\bar{\ell}}, = 0$ outside it ($g(\cdot)$ is the characteristic function of $S_{\bar{k}}^{\bar{\ell}}$). We see that: $g(\cdot) \in L^2(\mathbb{R}^n)$. Also, the scalar product:

$(\phi_j, g)_{L^2} = \int_{\mathbb{R}^n} \phi_j(\xi) g(\xi) d\xi$ equals $\int_{S_{\bar{k}}^{\bar{\ell}}} \phi_j(\xi) d\xi > \frac{1}{p_0} m(S_{\bar{k}}^{\bar{\ell}}) > 0$, $\forall j \geq \bar{k}$

which contradicts assumption i).

This establishes the "Proposition".

We are now ready for the

End of proof of Th. 7.1 We have seen already that, from (7.6), one derives

$$c_0 \hat{u}(\xi) = w(\xi) = \text{(weak } L^2 \text{ limit) of } (\mathcal{G}_{s,j_{k_\ell}} u)^\wedge(\xi).$$

In particular, for the $h \in L^2$ where (7.5) holds, we have also

$$\int_{\mathbb{R}^n} (\mathcal{G}_{s,j_{k_\ell}} u)^\wedge(\xi) \bar{\hat{h}}(\xi) d\xi \to \int_{\mathbb{R}^n} c_0 \hat{u}(\xi) \bar{\hat{h}}(\xi) d\xi$$

which is in contradiction with (7.5).

This way we proved the part A), that is the relation (7.1) for $u \in \mathcal{S}(\mathbb{R}^n)$.

B) We now establish (7.1) for all $u \in L^2(\mathbb{R}^n)$.

Let us use notation $\mathcal{G}_{s,j}$ for the operators in the left-hand side of (7.1).

Thus, we now prove:

$$\lim_{j \to \infty} \mathcal{G}_{s,j} u = c_0 u, \ \forall u \in L^2(\mathbb{R}^n), \text{ in weak-} L^2 \text{ sense.} \tag{7.9}$$

Note the uniform estimate:

$$\| \mathcal{G}_{s,j} u \|_{L^2} \leq \| \mathcal{G}(x, D) \|_{\mathcal{L}(H^s)} \cdot \| u \|_{L^2} \tag{7.10}$$

$\forall u \in L^2(\mathbb{R}^n), j = 1, 2, \ldots, s \in \mathbb{R}$.

(see Remark 2 above).

Because of (7.3) we know that

$$\lim_{j \to \infty} (\mathcal{G}_{s,j} u, h)_{L^2} = (c_0 u, h)_{L^2} \tag{7.11}$$

holds true, $\forall h \in L^2$ and $\forall u \in \mathcal{S}(\mathbb{R}^n)$.

Take now any $u \in L^2(\mathbb{R}^n)$, and fix also $h \in L^2(\mathbb{R}^n)$. For any $\mathcal{E} > 0$, there is $u_{\mathcal{E}} \in \mathcal{S}(\mathbb{R}^n)$, such that $\| u_{\mathcal{E}} - u \|_{L^2} < \frac{\mathcal{E}}{2(\|\mathcal{G}\|_{L(H^s)}\|h\| + |c_0|\|h\|)}$. We have also the simple relation

$$(\mathcal{G}_{s,j}u - c_0 u, h) =$$
$$(\mathcal{G}_{s,j}(u - u_{\mathcal{E}}), h)_{L^2} - (c_0(u - u_{\mathcal{E}}), h)_{L^2} + (\mathcal{G}_{s,j}u_{\mathcal{E}} - c_0 u_{\mathcal{E}}, h)_{L^2} \tag{7.12}$$

whence the estimates

$$|(\mathcal{G}_{s,j}u - c_0 u, h)| \leq \| \mathcal{G}(x, D) \|_{\mathcal{L}(\mathcal{H}^s} \| u - u_{\mathcal{E}} \| \, \| h \| + |c_0| \, \| u - u_{\mathcal{E}} \| \, \| h \| +$$
$$|(\mathcal{G}_{s,j}u_{\mathcal{E}} - c_0 u_{\mathcal{E}}, h)| < \frac{\mathcal{E}}{2} + (\mathcal{G}_{s,j}u_{\mathcal{E}} - c_0 u_{\mathcal{E}}, h)| \tag{7.13}$$

Using (7.11) we may find $\bar{j}(\mathcal{E})$ such that, $j \geq \bar{j}(\mathcal{E}) \Rightarrow |(\mathcal{G}_{s,j}u_{\mathcal{E}} - c_0 u_{\mathcal{E}}, h)| < \frac{\mathcal{E}}{2}$.
Altogether we obtain
$|(\mathcal{G}_{s,j}u, h)_{L^2} - (c_0 u, h)_{L^2}| < \mathcal{E}$ for $j \geq \bar{j}(\xi)$, with u and h fixed in L^2.
This proves part B) and the Theorem 7.1.

8.

In this section we shall discuss a (fundamental) result concerning the "true order" of the above considered operators $\mathcal{G}(x, D)$ and $G(x, D)$.

If we refer (again) to Ch. IV and see that assumptions (1.1)–(1.2)–(1.3) imply that *both* operators $\mathcal{G}(x, D)$ *and* $G(x, D)$ are linear continuous operators $H^s(\mathbb{R}^n) \to H^s(\mathbb{R}^n)$, for all $s \in \mathbb{R}$. Thus, taking $V = H^\infty(\mathbb{R}^n)$ (see Ch. I), we see that $0 \in$ "order set" of both $\mathcal{G}$ and G. Accordingly, the *true orders* of both $\mathcal{G}$ and G are ≤ 0. This can be strengthened. First we present

Theorem 8.1 *Under the assumption of Th. 7.1, the true order of the operator* $\mathcal{G}(x, D)$ *equals 0, unless* $\limsup_{|\xi| \to \infty} |g(x, \xi)| = 0 \forall x \in \mathbb{R}^n$

Proof We shall prove in fact the following:

$$\text{if } t.o(\mathcal{G}(x, D)) < 0, \text{ then } \lim_{|\xi| \to \infty} \sup |g(x, \xi)| = 0 \ \forall x \in \mathbb{R}^n. \tag{8.1}$$

In fact, if $t.o(\mathcal{G}(x, D)) = \ell < 0$, there is $\mathcal{E} > 0$ such that $-\mathcal{E} \in \mathcal{O}(\mathcal{G}(x, D))$. Then we have the estimates ($\forall s \in \mathbb{R}$)

$$\| \mathcal{G}(x, D)u \|_{H^s} \leq C_s \| u \|_{H^{s-\mathcal{E}}} \ \ \forall u \in H^\infty \text{ (hence } \forall u \in \mathcal{S}(\mathbb{R}^n) \text{) too).} \tag{8.2}$$

Let us assume now, reasoning by contradiction, that there exists $x_0 \in \mathbb{R}^n$, such that $\limsup_{|\xi|\to\infty} |g(x_0,\xi)| > 0$. This means that $\lim_{R\to\infty} (\sup_{|\xi|\geq R} |g(x_0,\xi)|) = \ell > 0$. Therefore, $\exists \bar{R}$ such that

$$\sup_{|\xi|\geq r} |g(x_0,\xi)| > \frac{\ell}{2} \text{ for } r \geq \bar{R} \tag{8.3}$$

Then, we first find $\xi_1, |\xi_1| \geq \bar{R}$ such that $|g(x_0,\xi_1)| > \frac{\ell}{2}$. Next, take $r > \max(\bar{R}+1, |\xi_1|)$, so that $\sup_{|\xi|\geq r} |g(x_0,\xi)| > \frac{\ell}{2}$ and accordingly, for some ξ_2 with $|\xi_2| > |\xi_1|$, we have $|g(x_0,\xi_2)| > \frac{\ell}{2}$ (also $|\xi_2| > \bar{R}+1$). Again, take $r > \max(|\xi_2|, \bar{R}+2)$, such that $\sup_{|\xi|\geq r} |g(x_0,\xi)| > \frac{\ell}{2}$ and then $\xi_3, |\xi_3| > |\xi_2|$ and $|\xi_3| > \bar{R}+2$, such that $|g(x_0,\xi_3)| > \frac{\ell}{2}$. Continuing in the same way, we now have a sequence $(\xi_p)_1^\infty$ in $\mathbb{R}^n$, where $|\xi_p| \geq \bar{R}$ and $|\xi_p| \to \infty$ (as $p \to \infty$), such that

$$|g(x_0,\xi_p)| > \frac{\ell}{2} \ \forall p = 1,2,\ldots \tag{8.4}$$

Consider now the sequence $(g(x_0,\xi_p))_{p=1}^\infty$ which is a bounded numerical sequence; hence, for a subsequence $(g(x_0,\xi_{p_j}))_{j=1}^\infty$ we have

$$\lim_{j\to\infty} g(x_0,\xi_{p_j}) = C_0 \text{ (a complex number)} \tag{8.5}$$

Furthermore, as $|g(x_0,\xi_{p_j})| > \frac{\ell}{2} > 0$, it follows, since also

$$|g(x_0,\xi_{p_j})| \to |C_0|, \text{ that } |C_0| \geq \frac{\ell}{2} > 0, \text{ so that } C_0 \neq 0. \tag{8.6}$$

We shall now use Th. 7.1. We get a subsequence of $(\xi_{p_j})_1^\infty$ – let us call it $(\xi_j^1)_1^\infty$ – such that $Q^s_{x_0,\xi_j^1} \mathcal{G}(x,D) P^s_{x_0,\xi_j^1} u \to C_0 u$, weakly in $L^2, \forall u \in \mathcal{S}(\mathbb{R}^n)$.

Consider the sequence of elements, $(P^s_{x_0,\xi_j^1} u)_{j=1}^\infty$, which belong to $\mathcal{S}(\mathbb{R}^n)(\forall j = 1,2,\ldots)$. Apply then (8.2) and obtain the estimates

$$\| \mathcal{G}(x,D) P^s_{x_0,\xi_j^1} u \|_{H^s} \leq C_s \| P^s_{x_0,\xi_j^1} u \|_{H^{s-\varepsilon}} \tag{8.7}$$

On the other hand, we have: $\| P^s_{x_0,\xi_j^1} u \|_{H^{s-\varepsilon}} = \| I_{-s} P^s_{x_0,\xi_j^1} u \|_{H^{s-\varepsilon}} = \| P^s_{x_0,\xi_j^1} u \|_{H^{-\varepsilon}}$ which $\to 0$ as $j \to \infty$ (Prop. 4.1).

Therefore we get:

$$\lim_{j\to\infty} \| \mathcal{G}(x,D)P^s_{x_0,\xi^1_j}u \|_{H^s} = 0 \tag{8.8}$$

and, accordingly, as any $Q^s_{x_0,\xi^1_j}$ belongs to $\mathcal{L}(H^s;H^0)$, that

$$\lim_{j\to\infty} \| Q^s_{x_0,\xi^1_j}\mathcal{G}(x,D)P^s_{x_0,\xi^1_j}u \|_{H^0} = 0,\ \forall u\in\mathcal{S}(\mathbb{R}^n),$$

so that the sequence $(\mathcal{G}_{s,j}(x,D)u)_1^\infty$ is strongly (and weakly) convergent to 0 while it also converges weakly to C_0u. Thus: $C_0u = 0 \forall u \in \mathcal{S}(\mathbb{R}^n)$ and $C_0 = 0$, which contradicts (8.6).

■

Next, we shall attempt to establish a similar result ($t.oG(x,D) = 0$) for the operator $G(x,D)$ associated to $g(x,\xi)$ through formula

$$G(x,D)U = \mathcal{F}^{-1}\left[(2\pi)^{-n/2}\int_{\mathbb{R}^n}\gamma(\xi-\eta,\xi)\hat{U}(\eta)d\eta + \tilde{g}(\xi)\hat{U}(\xi)\right],\ \forall U\in H^s(\mathbb{R}^n) \tag{8.9}$$

(see section 2, Ch. IV). We only need the previous result ($t.o\mathcal{G}(x,D) = 0$) together with a special estimate of the operator $G(x,D) - \mathcal{G}(x,D)$.

Again we note that the Fourier transform: $\mathcal{F}(G-\mathcal{G})U$ (where $U \in H^s$) is expressed by the integral

$$(2\pi)^{-n/2}\int_{\mathbb{R}^n}[\gamma(\xi-\eta,\xi)-\gamma(\xi-\eta,\eta)]\hat{U}(\eta)d\eta = W(\xi)\in L^1_{loc}(\mathbb{R}^n). \tag{8.10}$$

Next, we consider the expression

$$W_s(\xi) = (2\pi)^{-n/2}(1+|\xi|^2)^{s+1/2}\int_{\mathbb{R}^n}[\gamma(\xi-\eta,\xi)-\gamma(\xi-\eta,\eta)]\hat{U}(\eta)d\eta, \xi\in\mathbb{R}^n/\{o\} \tag{8.11}$$

which can be written as

$$W_s(\xi) = (2\pi)^{-n/2}\int_{\mathbb{R}^n}(1+|\xi|^2)^{s+1/2}(1+|\eta|^2)^{-s+1/2}[\gamma(\xi-\eta,\xi)-\gamma(\xi-\eta,\eta)]$$
$$(1+|\eta|^2)^{s+1/2}\hat{U}(\eta)d\eta$$

Using Peetre's inequality as well as assumption (1.5) in this Chapter, we obtain estimate:

$$|W_s(\xi)| \le (2\pi)^{-n/2}2^{|s+1|/2}\int_{\mathbb{R}^n}(1+|\xi-\eta|^2)^{|s+1|/2}$$

$$(1+|\xi-\eta|)\varphi(|\eta|)k(\xi-\eta)(1+|\eta|^2)^{s+1/2}|\hat{U}(\eta)|d\eta$$

$$(\forall\xi\in\mathbb{R}^n/\{o\}) \tag{8.12}$$

Let us obtain now a few simple estimates for $\varphi(t)$. We have: $|\varphi(t)| < \frac{1}{\sqrt{t}}$ for $t \geq \bar{t}$ – (1.4). Furthermore, note that: $t^2 > \frac{1}{2}(1+t^2)$ for $t > 1$, hence $t^{1/2} > (\frac{1}{2})^{1/4}(1+t^2)^{1/4}$ for $t > 1$, so that: $0 < \varphi(t) < C(1+t^2)^{-1/4}$ for $t \geq \max(\bar{t}, 1)$. Let also: $C_1 = \sup_{0\leq t\leq\max(\bar{t},1)} (1+t^2)^{1/4}\varphi(t)$. Then we get: $(1+t^2)^{1/4}\varphi(t) \leq \max(C, C_1) = C_2$ for all $t \geq 0$, that is: $\varphi(t) \leq C_2(1+t^2)^{-1/4}, \forall t \geq 0$. Note also that:

$$(1+|\lambda|) \leq \sqrt{2}(1+|\lambda|^2)^{1/2}, \forall\lambda\in\mathbb{R}^n. \tag{8.13}$$

Accordingly, introducing in (8.12) we obtain the inequality:

$$|W_s(\xi)| \leq C_s \int_{\mathbb{R}^n} (1+|\xi-\eta|^2)^{|s+1|/2+1/2}k(\xi-\eta)\cdot(1+|\eta|^2)^{-1/4+(s+1)/2}|\hat{U}(\eta)|d\eta$$

$$,\forall\xi\in\mathbb{R}^n/\{o\} \tag{8.14}$$

Next, in the right-hand side of (8.14) we have the convolution between the function $(1+|\lambda|^2)^{|s+1|/2+1/2}k(\lambda)$ (which belongs to $L^1(\mathbb{R}^n)\forall s\in\mathbb{R}$) – see (1.3) – and the function $(1+|\eta|^2)^{s/2+1/4}|\hat{U}(\eta)|$ (which belongs to $L^2(\mathbb{R}^n)\forall s\in\mathbb{R}$ if we assume *(as we do here)* that $U\in\mathcal{S}(\mathbb{R}^n)$ (in fact $U\in H^{s+(1/2)}$ suffices).

Therefore (again by Young's theorem) we find the upper estimate:

$$\| W_s(\cdot) \|_{L^2} \leq C_{k,s} \| (1+|\eta|^2)^{s/2+1/4}|\hat{U}(\eta) \|_{L^2}$$

which means that

$$\| (G-\mathcal{G})U \|_{H^{s+1}} \leq C_{k,s}\left(\int_{\mathbb{R}^n}(1+|\eta|^2)^{s+(1/2)}|\hat{U}(\eta)|^2 d\eta\right)^{1/2} = C_{k,s} \| U \|_{H^{s+(1/2)}} \tag{8.15}$$

and this $\forall s\in\mathbb{R}, \forall U\in\mathcal{S}(\mathbb{R}^n)$ (or for $U\in H^\infty(\mathbb{R}^n)$).

We are now ready for the statement of

Theorem 8.2 *Under the assumptions of Th. 8.1, the true order of the operator $G(x,D)$ equals 0, unless $\limsup_{|\xi|\to\infty}|g(x,\xi)| = 0\forall x\in\mathbb{R}^n$.*

Proof Let us again assume: $t.o(G(x,D)) < 0$. Then, $\exists\mathcal{E} > 0$, such that

$$\| G(x,D)U \|_{H^s} \leq C_s \| U \|_{H^{s-\mathcal{E}}}, \forall U\in\mathcal{S}(\mathbb{R}^n) \tag{8.16}$$

Let us write now: $\mathcal{G}(x,D) = \mathcal{G}(x,D) - G(x,D) + G(x,D)$. It follows:

$$\| \mathcal{G}(x,D)U \|_{H^s} \leq \| [\mathcal{G}(x,D) - G(x,D)]U \|_{H^s} + C_s \| U \|_{H^{s-\mathcal{E}}}, \forall U \in \mathcal{S}(\mathbb{R}^n) \quad (8.17)$$

If we use (8.15) we derive accordingly

$$\| \mathcal{G}(x,D)U \|_{H^s} \leq C_s \| U \|_{H^{s-1/2}} + C_s \| U \|_{H^{s-\mathcal{E}}}, \forall U \in \mathcal{S}(\mathbb{R}^n), \ \forall s \in \mathbb{R}$$

thus $\| \mathcal{G}(x,D)U \|_{H^s} \leq C_s \| U \|_{H^{s-\sigma}}$ where $\sigma = \mathcal{E}$ if $0 < \mathcal{E} < \frac{1}{2}$ and $\sigma = \frac{1}{2}$ if $\mathcal{E} \geq \frac{1}{2} (\forall u \in \mathcal{S}(\mathbb{R}^n))$.

Then we are in the case of Th. 8.1, so that, again, $\limsup_{|\xi| \to \infty} |g(x,\xi)| = 0, \forall x \in \mathbb{R}^n$.

Example (Kohn–Nirenberg C^∞ and 0-homogeneous symbols. With the notation in Ch. VI, we now have symbols $a(x,\xi) = a(\infty,\xi) + a'(x,\xi)$ where $a'(x,\xi) = (2\pi)^{-n/2} \int_{\mathbb{R}^n} e^{i<x,\lambda>} \tilde{a}'(\lambda,\xi) d\lambda$, so that

$$a(x,\xi) = (2\pi)^{-n/2} \int_{\mathbb{R}^n} e^{i<x,\lambda>} \tilde{a}'(\lambda,\xi) d\lambda + a(\infty,\xi), \ \forall (x,\xi) \in \mathbb{R}^{\times}\mathbb{R}^n/\{o\} \quad (8.18)$$

Here the (partial) Fourier transform $\tilde{a}'(\lambda,\xi)$ is continuous (hence measurable) in $\mathbb{R}^n \times \mathbb{R}^n/\{o\}$ (see Ch. VI, 2); also, the function $\lambda \to \tilde{a}'(\lambda,\xi)$ is measurable, $\forall \xi \in \mathbb{R}^n/\{o\}$.

Furthermore, the function $a(\infty,\xi)$ is continuous and bounded in $\mathbb{R}^n/\{o\}$.

Next, we prove the estimate (1.6), that is

$$|a(\infty,\xi) - a(\infty,\eta)| \leq (1 + |\xi - \eta|)\varphi_1(|\xi|), \ \forall \xi,\eta \in \mathbb{R}^n/\{o\}$$

(with a convenient function $\varphi_1(\cdot)$ satisfying (1.4)).

We use (2.25) in Ch. VI: $\frac{|\xi-\eta|}{|\xi|+|\eta|} \leq (1 + |\xi - \eta|)(1 + |\eta|)^{-1}$, for $\xi,\eta \in \mathbb{R}^n/\{o\}$ and (1.6) in Ch. VI: $|a(\infty,\xi) - a(\infty,\eta)| \leq C|\xi - \eta|(|\xi| + |\eta|)^{-1}, \xi,\eta \in \mathbb{R}^n/\{o\}$ to get:

$|a(\infty,\xi) - a(\infty,\eta)| \leq C(1 + |\eta|)^{-1}(1 + |\xi - \eta|)$, or also, changing ξ with η:

$$|a(\infty,\xi) - a(\infty,\eta)| \leq C(1 + |\xi|)^{-1}(1 + |\xi - \eta|), \ \forall \xi,\eta \in \mathbb{R}^n/\{o\}$$

which is (1.6) with $\varphi_1(t) = \frac{C}{1+t}, t \geq 0$.

Next we make appeal to the *Remark* at the end of Ch. VII. We found there a function $k(\lambda)$ such that $|\tilde{a}'(\lambda,\xi)| \leq k(\lambda), \forall\lambda \in \mathbb{R}^n, \xi \in \mathbb{R}^n/\{o\}$, where $k(\lambda).(1+|\lambda|^2)^s \in L^1(\mathbb{R}^n)\forall s \in \mathbb{R}$, and also

$$|\tilde{a}'(\lambda,\xi) - \tilde{a}'(\lambda,\eta)| \leq (1+|\xi-\eta|)k(\lambda)(1+|\xi|^2)^{-1/2}, \ \forall\lambda \in \mathbb{R}^n, \xi, \eta \in \mathbb{R}^n/\{o\}.$$

This gives our condition (1.5) with $\varphi_2(t) = \frac{1}{\sqrt{1+t^2}}$.

Therefore, taking $\varphi(t) = \max(\varphi_1(t), \varphi_2(t)), t \geq 0$, we get a continuous function such that (1.4)–(1.5)–(1.6) are all verified – for $\tilde{a}'(\lambda,\xi)$ and $a(\infty,\xi) = \tilde{g}(\xi)$.

We can apply Th. 8.1 and 8.2 and obtain that $t.oA(x,D) = t.o\mathcal{A}(x,D) = 0$ unless $\limsup_{|\xi|\to\infty} |a(x,\xi)| = 0, \forall x \in \mathbb{R}^n$. Using homogeneity of $a(x,\xi)$ we get: $a(x,\xi) = a(x, \frac{\xi}{|\xi|}), \forall\xi \in \mathbb{R}^n/\{o\}$. Therefore, if $r > 1$, $\sup_{|\xi|>r} |a(x,\xi)| = \sup_{|\xi|=1} |a(x,\xi)|$.

Therefore, $\lim_{r\to\infty} \sup_{|\xi|>r} |a(x,\xi)| = \limsup_{|\xi|\to\infty} |a(x,\xi)| = \sup_{|\xi|=1} |a(x,\xi)|$.

Consequently, $\limsup_{|\xi|\to\infty} |a(x,\xi)| = 0 \forall x \in \mathbb{R}^n \iff \sup_{|\xi|=1} |a(x,\xi)| = 0, \forall x \in \mathbb{R}^n$ which means that $a(x,\xi) = 0\forall(x,\xi) \in \mathbb{R}^n \times \mathbb{R}^n/\{o\}$.

■

References

[1] Andreotti, A. and Spagnolo, S. (1966-67) *Operatori pseudodifferenziali*, Università degli Studi di Pisa.

[2] Cordes, H.O. (1979) *Elliptic Pseudo-Differential Operators, An abstract theory*, Springer-Verlag, Berlin–Heidelberg–New York.

[3] Dunford, N. and Schwartz J.T. (1958) *Linear Operators, Part I : General Theory*, Interscience Publishers, Inc., New York.

[4] Friedrichs, K.O. (1970) Pseudo-Differential Operators (An introduction), *Courant Institute of Mathematical Sciences*, New York University.

[5] Kohn, J.J. and Nirenberg, L. (1965) An algebra of pseudo-differential operators, *Comm. Pure Appl. Math.*, **18**, 267-305.

[6] Petersen, B.E. (1983) *Introduction to the Fourier transform and Pseudo-Differential Operators*, Pitman Advanced Publishing Program, Boston–London–Melbourne.

[7] Royden, H.L. (1968) *Real Analysis* (2nd ed.), MacMillan Publishing Co. Inc., New York.

[8] Yosida, K. (1978) *Functional Analysis* (5th ed.), Springer-Verlag, Berlin.

[9] Zaidman, S. (1991) Distributions and Pseudo-Differential Operators, *Pitman RNM*, 248, Longman Scientific and Technical.

[10] Zaidman, S. (1972) Pseudo-Differential Operators, *Annali di Matematica pura ed applicata*, (IV), Vol. XCII, 345-399.

[11] Zaidman, S. (1983) Asymptotic expansions of linear operators in some vector spaces, *Ann. Sc. Math. Quebec*, vol.7, no.2, 203-208.

[12] Zaidman, S. (1983) Compacité de l'opérateur $A(x, D) - \mathcal{A}(x, D)$ dans l'espace $\mathcal{F}^{-1}(L^1(\mathbb{R}^n))$, *Ann. Sc. Math. Quebec*, vol.7, no.2, 208-218.

[13] Zaidman, S. (1986) Order and true order of linear operators in some vector spaces, *Bull. Math. S.S.M. R.S.R.*, vol.30 (78)2, 179-186.

[14] Zaidman, S. (1993) A lower estimate for the norm of a pseudodifferential operator modulo compact operators, *Proc. Intern. Symposium on Generalized Functions and Their Applications*, Plenum Press, 293-297.

[15] Zaidman, S. (1970) Certaines Classes d'Opérateurs Pseudo-Differentiels, *J. Math. Anal. Appl.*, vol.30, 522-563.

Index of symbols

E	(Banach space)
$\{E^s\}$	(scale of Banach spaces)
K	(real or complex field)
$\| \ \|_{E^s}$	(norm in the space E^s, for $s \in \mathbb{R}$)
$H^s(\mathbb{R}^n)$	(Soboleff space)
$\mathbb{N}$	(natural numbers)
$\mathbb{C}$	(complex field)
$\mathcal{B}_{1,s}(\mathbb{R}^n)$	$(T \in \mathcal{S}'(\mathbb{R}^n), (1+\|\xi\|^2)^{s/2}\hat{T}(\xi) \in L^1(\mathbb{R}^n))$
E^∞	(intersection of all the E^s, for $s \in \mathbb{R}$)
H^∞	(intersection of all the H^s, for $s \in \mathbb{R}$)
V	(vector subspace of E^∞)
$\mathcal{S}(\mathbb{R}^n)$	(space of $C^\infty(\mathbb{R}^n)$ and rapidly decreasing functions)
$\mathcal{B}_{1,\infty}(\mathbb{R}^n)$	(the intersection of all the $\mathcal{B}_{1,s}(\mathbb{R}^n), s \in \mathbb{R}$)
$\mathbb{R}$	(the real line)
$\mathcal{L}in(V)$	(vector space of all linear operators $V \to V$, over K)
$\mathcal{O}(L)$	(the order set of $L \in \mathcal{L}in(V)$)
$\mathcal{L}(E^s, E^\sigma)$	(space of all linear continuous operators, $E^s \to E^\sigma$)
$\mathcal{P}(\mathbb{R}) = 2^{\mathbb{R}}$	(class of all the subsets of $\mathbb{R}$)
$\mathbb{R}^n$	(the n-dimensional euclidean space)
$L^p(\mathbb{R}^n), p \geq 1$	(the usual space of Lebesgue p-integrable functions)
$\hat{u}, \mathcal{F}u$	(Fourier transformation of u)
$\mathcal{F}^{-1}$	(inverse Fourier transformation)
$\mathcal{S}'(\mathbb{R}^n)$	(space of temperate distributions over $\mathbb{R}^n$)
$\varphi(D) = \mathcal{F}^{-1}\mathcal{M}_{\varphi(\cdot)}\mathcal{F}$	(Friedrichs pseudo-differential operator associated to the function $\varphi(\xi)$)
$\| \ \|_{H^s}$	(norm in the space $H^s(\mathbb{R}^n)$)
$\|\xi\|^2 = \xi_1^2 + \dots \xi_n^2$	(square of length in $\mathbb{R}^n$ of the vector $\xi = (\xi_1, \dots, \xi_n)$)
$p(x,\xi)$	(symbol of pseudo-differential operators)
$\tilde{S}^r$	(class of symbols $p(x,\xi) \in C^\infty(\mathbb{R}^n \times \mathbb{R}^n)$ such that $(1+\|x\|^2)^\ell \|\partial_\xi^\alpha \partial_x^\beta p(x,\xi)\| \leq C_{\alpha,\beta,\ell}(1+\|\xi\|)^{r-\|\alpha\|}$
$\mathcal{P}(x,D)$	(pseudo-differential operator associated with $p(x,\xi)$)
$< x, \xi >$	$(\sum_{i=1}^n x_i\xi_i$, scalar product)

∂_x^α	(partial derivative operator: $\partial_1^{\alpha_1}\partial_2^{\alpha_2}\ldots\partial_n^{\alpha_n} = \frac{\partial^{\alpha_1+\alpha_2+\ldots\alpha_n}}{\partial x_1^{\alpha_1}\partial x_2^{\alpha_2}\ldots\partial x_n^{\alpha_n}}$)
$\binom{\alpha}{\beta}$	($\frac{\alpha!}{\beta!(\alpha-\beta)!}$, binomial coefficients; $\alpha! = \alpha_1!\alpha_2!\ldots\alpha_n!$)
$\beta \le \alpha$	(for multi indexes $\alpha, \beta \in \underline{\mathbb{N}}^n$, where $\underline{\mathbb{N}} = (0,1,2,\ldots)$; $\beta_i \le \alpha_i, i = 1,2,\ldots n$)
i	(imaginary unit = $\sqrt{-1}$)
ξ^α	($\xi_1^{\alpha_1}\xi_2^{\alpha_2}\ldots\xi_n^{\alpha_n}$, where $\xi = (\xi_1\ldots\xi_n), \alpha = (\alpha_1,\ldots\alpha_n) \in \underline{\mathbb{N}}^n$)
$d\xi$	(element of volume in $\mathbb{R}^n : d\xi = d\xi_1 d\xi_2 \ldots d\xi_n$)
$\alpha - \beta$	(difference vector: $(\alpha_1 - \beta_1, \alpha_2 - \beta_2, \ldots \alpha_n - \beta_n)$)
$\lvert\gamma\rvert$	(length of multi-index $\gamma = (\gamma_1, \gamma_2 \ldots \gamma_n) : \lvert\gamma\rvert = \sum_{i=1}^{n} \gamma_i$)
$\mathcal{L}in_{\mathcal{O}\neq\phi}(V)$	(space of linear operators, $V \to V$ with non-empty order set)
$\mathcal{L}in_{\mathcal{O}=\mathbb{R}}(V)$	(space of linear operators, $V \to V$, with order set $= \mathbb{R}$)
$t.o(L)$	(true order of the linear operator L)
$\mathcal{L}_{-\infty}$	(another notation for $\mathcal{L}in_{\mathcal{O}=\mathbb{R}}(V)$)
$\mathcal{M}_{\varphi(\cdot)}$	(multiplication operator by the function $\varphi(\cdot)$)
$\mathbb{R}^+$	(open half-line $]0, +\infty[$)
$C_0^\infty(\mathbb{R}^n)$	(infinitely differentiable functions, $\mathbb{R}^n \to \mathbb{C}$, vanishing outside compact subsets of $\mathbb{R}^n$)
$A + B$	(vector sum of subsets in a vector space)
Θ	(the null-operator, $V \to V$)
$\sim$	(asymptotic equivalence)
$\sum_{j=0}^{\infty} A_j$	(asymptotic series of operators A_j in $\mathcal{L}in(V)$)
$\overset{p}{\sim}$	(p-equivalence)
$A \overset{p}{\sim} B$	(the operators A, B are p-equivalent)
$\sum_{j=0}^{k}$	(ordinary finite sum of operators A_j in $\mathcal{L}in(V)$)
$\overset{f}{\sim}$	(finite expansion)
D_j	(partial differential operator $\frac{1}{\sqrt{-1}}\frac{\partial}{\partial x_j}, j = 1,2,\ldots n$)
D^α	(operator $D_1^{\alpha_1}D_2^{\alpha_2}\ldots D_n^{\alpha_n}$ for $\alpha = (\alpha_1, \alpha_2 \ldots \alpha_n) \in \underline{\mathbb{N}}^n$)
$C^m(\mathbb{R}^n), C^\infty(\mathbb{R}^m)$	(vector space of functions $\varphi(\cdot), \mathbb{R}^n \to K$ such that $\partial^\alpha\varphi \in C(\mathbb{R}^n)$, for $\lvert\alpha\rvert \le m$ (or for all $\alpha \in \underline{\mathbb{N}}^n$))
$\hat{a} * \hat{U}$	(convolution between $L^1(\mathbb{R}^n)$-functions, $\hat{a}(\xi)$ and $\hat{U}(\xi)$)
$\mathcal{L}(\mathcal{B}_{1,0}, \mathcal{B}_{1,0})$	(space of all linear continuous operators, $\mathcal{B}_{1,0} \to \mathcal{B}_{1,0}$)
$ess.sup_{\mathbb{R}^n}$	(essential supremum on $\mathbb{R}^n$)
$\mathbb{R}^n \times \mathbb{R}^n$	(cartesian product of two copies of $\mathbb{R}^n$)

grad	(gradient $= (\frac{\partial}{\partial\xi_1}, \ldots \frac{\partial}{\partial\xi_n})$)
$a.e$	(almost-everywhere)
$\mathcal{G}_{ix_j}(x,D)$	(pseudo-differential operator associated to the symbol $\frac{1}{i}\frac{\partial g}{\partial x_j}$ (where $g = g(x,\xi)$ is the symbol of $\mathcal{G}(x,D)$))
$\mathcal{F}^{-1}(L^p(\mathbb{R}^n))$	(space of temperate distributions with Fourier transform in $L^p(\mathbb{R}^n)$)
$G(x,D), \mathcal{G}(x,D)$	(couple of pseudo-differential operators associated with symbol $g(x,\xi)$)
$L^1_{loc}(\mathbb{R}^n)$	(functions $\varphi(\cdot), \mathbb{R}^n \to \mathbb{C}$, such that $\int_K \lvert\varphi(\xi)\rvert d\xi < +\infty$ for all K, compact set in $\mathbb{R}^n$)
$T_{K(\cdot)}$	((temperate) distribution associated with function $K(\xi)$)
$(I-\Delta), (I-\Delta)^m$	(differential operator $I - \sum_{i=1}^{n} \frac{\partial^2}{\partial x_i^2}$; its mth iterate)
$\bigcup_1^\infty \mathcal{E}_p$	(countable union of the sets $\mathcal{E}_p$)
$m\mathcal{E}$	(Lebesgue measure of the set $\mathcal{E}$ in $\mathbb{R}^n$)
$\mathbb{R}^n/\mathcal{E}$	(difference set between $\mathbb{R}^n$ and set $\mathcal{E}$)
$S(x_0,r)$	(ball, $x \in \mathbb{R}^n, \lVert x - x_0 \rVert < r$)
$A(x,D), \mathcal{A}(x,D)$	(couple of pseudo-differential operators associated with symbol $a(x,\xi)$)
$\mathbb{R}^n/\{o\}$	(open set of $x \in \mathbb{R}^n, \lVert x \rVert > 0$)
∂_ξ^β	(partial differential operator $\partial_1^{\beta_1}\partial_2^{\beta_2}\ldots\partial_n^{\beta_n}$)
$\underline{\mathbb{N}}$	(numbers $0,1,2,\ldots$)
$\underline{\mathbb{N}}^n$	(multi-indexes $\{\alpha_1,\alpha_2\ldots\alpha_n\}$, where $\alpha_j \in \underline{\mathbb{N}}\ \forall j = 1,2,\ldots n$)
$<\zeta,\mu>$	(scalar product $= \sum_{i=1}^{n} \zeta_i\mu_i$ of vectors ζ,μ in $\mathbb{R}^n$)
$[\zeta,\mu]$	(segment joining vectors ζ and μ in $\mathbb{R}^n$: $\{t\zeta + (1-t)\mu, 0 \leq t \leq 1\}$)
$\tilde{a}'(\lambda,\xi)$	(partial Fourier transform of the function $a'(x,\xi)$ with respect to the x-variable)
$a(x,\xi) = a(\infty,\xi) + a'(x,\xi)$	(Kohn-Nirenberg symbols; O-homogeneous and C^∞ in $\mathbb{R}^n \times \mathbb{R}^n/\{o\}$)
$\mathbb{R}^n \times S_1$	(cartesian product between $\mathbb{R}^n_x$ and the unit ball in $\mathbb{R}^n_\xi$)
$\mathbb{R}^n_x, \mathbb{R}^n_\lambda$	(notation for the $\mathbb{R}^n$-space)
$A_a(x,D), \mathcal{A}_a(x,D)$	(couple of pseudo-differential operators associated with symbol $a(x,\xi)$)

$\bar{a}(x,\xi)$	(complex-conjugate of the symbol $a(x,\xi)$)
$\overset{\simeq}{a'}$	(complex-conjugate of function $a'(x,\xi)$ followed by the x-partial Fourier transform)
$\bar{\hat{v}}$	(complex-conjugate of the Fourier transform of v)
$\bar{\simeq}$	(complex-conjugation followed by x-partial Fourier transform, then again by complex-conjugation)
Ω	(bounded set in $\mathcal{B}_{1,0}$)
$\sup\limits_{x\in\mathbb{R}^n} \overline{\lim\limits_{\lvert\xi\rvert\to\infty}} \lvert a(x,\xi)\rvert$	(supremum with respect to $x \in \mathbb{R}^n$ of the function $\varphi(x) = \limsup_{\lvert\xi\rvert\to\infty} \lvert a(x,\xi)\rvert$)
$C_b(\mathbb{R}^n)$	(space of bounded continuous functions, $\mathbb{R}^n \to \mathbb{C}$)
$P_{x_0,\xi_0}, Q_{x_0,\xi_0}$	(couple of linear operators defined on functions $\mathbb{R}^n \to \mathbb{C}$)
I_s	(mapping, $\mathcal{S}(\mathbb{R}^n) \to \mathcal{S}(\mathbb{R}^n)$, depending on $s \in \mathbb{R}$)
$(I-\Delta)^{s/2}$	(pseudo-differential operator I_s)
$P^s_{x_0,\xi_0}$	(operator $I_{-s}P_{x_0,\xi_0}$; $s \in \mathbb{R}$)
$Q^s_{x_0,\xi_0}$	(operator $Q_{x_0,\xi_0}I_s$; $s \in \mathbb{R}$)
I_s	(mapping $(I-\Delta)^{s/2}; \mathcal{S}'(\mathbb{R}^n) \to \mathcal{S}'(\mathbb{R}^n)$)
$\mathcal{G}_s(x,D)$	(product of mappings : $Q^s_{x_0,\xi_0}\mathcal{G}(x,D)P^s_{x_0,\xi_0}$)
R_{x_0,ξ_0}	(operator on functions $\mathbb{R}^n \to \mathbb{C}$)
$\mathcal{R}e, \mathcal{I}m$	(real part, imaginary part)
$\limsup\limits_{\lvert\xi\rvert\to\infty} \lvert g(x,\xi)\rvert$	(the limit: $\lim\limits_{R\to\infty} \sup\limits_{\lvert\xi\rvert\geq R} \lvert g(x,\xi)\rvert$)

Subject index

algebra (of linear operators) 2
asymptotic expansion (of linear operators) 17
bounded measurable functions 3
constant scale of Banach spaces 8
convolution (in $L^1(\mathbb{R}^n)$) 29
couple of pseudo-differential operators $G(x,D), \mathcal{G}(x,D)$ or $A(x,D), \mathcal{A}(x,D)$, in spaces $\mathcal{B}_{1,s}(\mathbb{R}^n), \mathcal{F}^{-1}(L^p), H^s(\mathbb{R}^n)$ 31, 34, 40, 43, 60
difference pseudo-differential operator $A - \mathcal{A}$ 63
order of 107
compactness of 67
field K 1
finite expansion (of linear operators) 25
Fourier transformation 4
Friedrichs operator 4, 9
Gohberg's Lemma 79
ideal (right and left) 7
identity mapping 7
inverse Fourier transform 3
Kohn-Nirenberg C^∞-homogeneous symbols 50
pseudo-differential operators 60
linear mapping, operator 1, 2
subspace, space 1, 2
Lipschitz condition on the unit ball of $\mathbb{R}^n$ 52
multiplication operator 28
in $H^s(\mathbb{R}^n)$ 44
norm of operator $\beta(D)$ 30
operator (linear continuous) 34

operator norm 3, 29
operator $Q_{x_0,\xi_0}\mathcal{G}(x,D)P_{x_0,\xi_0}$ 83
$P_{x_0,\xi_0}, Q_{x_0,\xi_0}$ 81
$I_s = (I - \Delta)^{s/2}$ 88
$P^s_{x_0,\xi_0}, Q^s_{x_0,\xi_0}$ 89
$\mathcal{G}_s(x,D) = Q^s_{x_0,\xi_0}\mathcal{G}(x,D)P^s_{x_0,\xi_0}$ 91
R_{x_0,ξ_0} 92
order of linear operator 2
set 3
partial differential operator 27
representation formula of 28
partial Fourier transformation 53, 54
Peetre's inequality (applied to ...) 42
p-equivalence (of linear operators) 18
polynomial symbol 28
pseudo-differential operators 4, 5
symbols of 34, 41, 42
relative compactness of sets in $\mathcal{B}_{1,0}(\mathbb{R}^n)$ 68
scale of Banach spaces 1
Schwartz $\mathcal{S}(\mathbb{R}^n)$-space 2
scale of Hilbert spaces 1
Soboleff $H^s(\mathbb{R}^n)$-space 42
temperate distributions 1
true order 2, 3
of operators $G(x,D), \mathcal{G}(x,D)$ 79, 104, 107